SLOW DOWN

SLOW DOWN

THE DEGROWTH
MANIFESTO

Kōhei Saitō

Translated by Brian Bergstrom

ASTRA HOUSE ∧ NEW YORK

Astra House
A Division of Astra Publishing House
astrahouse.com
Printed in the United States of America

Library of Congress Cataloging-in-Publication Data

Names: Kōhei Saitō, 1987– author.
Title: Slow down : the degrowth manifesto / written by Kōhei Saitō;
translated by Brian Bergstrom.
Other titles: Hitoshinsei no "Shihonron." English
Description: First edition. | New York : Astra House, [2024] | Includes
bibliographical references. | Summary: "In SLOW DOWN, Kōhei Saitō
delivers a bold and urgent call for a return to Marxism in order to stop
climate change. Here he argues that by returning to a system of social
ownership, we can restore abundance and focus on those activities that
are essential for human life, effectively reversing climate change and
saving the planet"— Provided by publisher.
Identifiers: LCCN 2023024298 (print) | LCCN 2023024299 (ebook) |
ISBN 9781662602368 (hardcover) | ISBN 9781662602351 (ebook)
Subjects: LCSH: Marxian economics. | Environmental economics. |
Marx, Karl, 1818-1883. | Capital.
Classification: LCC HB97.5 .S2317713 2024 (print) | LCC HB97.5 (ebook) |
DDC 335.4/12—dc23/eng/20230706
LC record available at https://lccn.loc.gov/2023024298
LC ebook record available at https://lccn.loc.gov/2023024299

First edition

10 9 8 7 6 5 4 3 2 1

Design by Input Data Services, Bridgwater, Somerset
The text is set in Minion Pro.
The titles are set in Minion Pro.

Contents

Preface to the English Edition

Not long ago I caught the train to Suzuka City, Mie Prefecture, about three hours west of Tokyo, to give a public lecture. Suzuka City boasts Honda's main automobile factory and is also known for the Suzuka Circuit, used for the Formula One World Championship, so the subject of the lecture made me a little anxious. Honda had recently announced its retreat from F1 racing to focus on greening the auto industry, but I was there to demolish the myth that electric vehicles would save the planet.

It was, as you might imagine, a tough crowd. After the exhausting lecture, as I was waiting for a train, a man came up to me. Looking a little nervous, he said he'd come from Tokyo just to hear me talk and that he had a question he wanted to ask. I imagined that he must be a climate activist of some kind, but to my surprise he introduced himself as the owner of a small rubber trading company, which in fact did business with a giant manufacturer. He told me that after reading my book, he could no longer tolerate his "stupid" business. It wasn't simply that he would always be at the mercy of bigger companies. He could no longer live with the fact that its products were helping to destroy the planet. At last he came to his question: What should he do with his rubber business?

I could not decide its fate for him during a ten-minute conversation. But actually, the answer to his question was in my book as well as in my lecture. Perhaps he had overlooked it or wanted another one that involved less change. My answer was that he should sign his business over to its employees. Since capitalism is the ultimate cause of climate breakdown, it is necessary to transition to a steady-state economy. All companies therefore need to become cooperatives or cease trading.

This encounter is quite representative of how even those who are successful and wealthy do not believe in the future prosperity of capitalism and are strongly attracted to new radical ideas. This guy is not an exception. In September 2020, a Japanese publisher, Shueisha, published this book under its original title, *Capital in the Anthropocene*. Even I thought my ideas were too radical to find much of an audience. Who would read a book on "degrowth communism" written by a basically unknown scholar of political thought in the Marxist tradition? But I was utterly wrong about that. It sold half a million copies.

Surely I owe some of this success to the fact that the book came out in the middle of the global pandemic, which meant that its message resonated with wider social discontentments and anxieties. Japanese society still suffers, like many others, from two of the main contradictions of capitalism. On the one hand, the pandemic increased economic inequality, which was painfully visible in the form of lines of unemployed workers and single mothers in search of food rations. Neoliberal reforms had cut the budgets of public health centers and infectious disease services, threatening the final collapse of the healthcare system. It was hard to ignore the fact that the capitalist system just wasn't delivering what society needed. At the same time, the pandemic showed what a severe ecological burden our daily existence imposed on the planet. We saw how our way of life increased the vulnerability of pretty much all living things to deadly anomalies and worrying precedents. We began to suspect that such developments may in fact be typical of the Anthropocene, the geological epoch in which the surface of the planet is entirely covered with the traces of human economic activity. In a threatening time, my book offered explanations.

There may have been more local factors behind its success. It begins with a chapter called "SDGs Are the Opiate of the Masses." In the face of the economic and ecological crisis of neoliberal capitalism, the UN's Sustainable Development Goals had become extremely popular in Japan, promoted by companies, politicians, and NGOs. But despite the extreme popularity of the concept, general interest in ecology, gender equality, and human rights remained quite superficial, such that extremely ineffective actions like eating broccoli stalks to reduce food waste and finishing every drink to reduce the use of PET bottles were being propagated in the media as "sustainable actions." In a seminar I gave, I even met a businessman wearing an SDGs badge on his jacket who apparently didn't know what the letters stood for. It seemed that no one dared to criticize the "sacred" concept of the SDGs. My book addressed this hypocrisy and accordingly won support among Japanese readers who do genuinely care about the environment and social justice. To my surprise, the phrase "SDGs are the opiate of the masses" actually went viral on Japanese social networking sites.

This level of absurdity may not exist outside Japan, but greenwashing is everywhere. It is popular precisely because it assures us that we do not have to change our current way of life, even though this way of life is based on the exploitation of other social groups and the destruction of natural environments in other regions. In this sense, an optimistic belief in green technologies and green growth may be nothing more than a ploy to buy time for capitalism. Far from being an encouraging development, the popularity of SDGs and green-growth jargon is yet another problem we need to solve.

Yet there are signs that the tide may turn. In recent years, advocates of degrowth have developed powerful critiques of capitalism, and there are important works in English on the subject, such as Jason Hickel's *Less Is More* and Tim Jackson's *Prosperity without Growth*, to which my own thinking is greatly indebted. The recent popularity of the degrowth concept more generally is also quite understandable given the repeated failure of attempts to make green capitalism work. In addition, we have now seen what can be done on short notice. During the pandemic the state pulled an emergency brake on capitalism, limiting economic activities and intervening in the market. Until that point, we didn't know that this economic system even had an emergency brake. It is in no way to minimize its tragic consequences to note that the slowdown in the consumerist way of life also created a space to think about the legitimacy of neoliberal capitalism, with its endless cycle of overproduction and overconsumption.

Proponents of degrowth are often ambivalent about the need to transcend capitalism. I am not ambivalent. In my opinion, degrowth must clarify its critical position *against* capitalism. This is why my book calls for degrowth *communism*. Stationary capitalism is *contradictio in adjecto*, as Joseph Schumpeter pointed out long ago. I am of course familiar with the standard objections that degrowth is impossible and communism is a nightmare. I cannot answer them in this short preface. As you read my book, I hope you will be convinced that green capitalism is a myth and that the future is indeed degrowth communism. Don't worry, it won't be a repeat of the old Marxist dogmas. I was born in 1987, so I never got to experience so-called actually existing socialism. That might seem like a disadvantage, but it offered one surprising benefit. Because I didn't reflexively impose Soviet history onto Marx's thought, my research into the vast corpus of his unpublished writings was able to uncover an entirely new aspect of his vision of the postcapitalist world, one that was perfectly adapted to the Anthropocene. Instead of the undemocratic state socialism controlled by the state bureaucrats, a more democratic, egalitarian, and sustainable vision of a new steady-state economy proves compatible with Marx's vision of the future society.

These "new" ideas of Marx's proved to be relevant amid Japan's crisis, especially to those who were passionate about exploring a new way of life. And there turned out to be more of these people than we thought. One of the clearest proofs is that in June 2022, my friend Satoko Kishimoto won the election for mayor in Tokyo's Suginami Ward. Although she worked for many years in Belgium for the international NGO Transnational Institute, she came back to Japan to run for office. She had no previous political experience and no backing from labor unions, but she campaigned on implicitly late-Marxian ideas of municipalism and the commons. She beat the Liberal Democratic Party (LDP) candidate by just 187 votes. This was a big surprise to many, including me, but it clearly shows that citizens are not indifferent to what might sometimes seem like rather academic concerns with ecology, feminism, and socialism. Voters do want a more egalitarian, sustainable, and just society.

Here's just one more recent development in Japanese politics that may be significant in light of the concerns of this book. Facing growing discontent during the COVID-19 lockdowns, the new prime minister Fumio Kishida put forward "new capitalism" as his key political vision and set up a Council of New Form of Capitalism Realization. I was not invited, perhaps in part because I am on record as opposing new forms of capitalism realization. Kishida's New Capitalism was soon watered down in the face of a rapid fall of stock price. But a notable change is nevertheless discernible here. The prime minister explicitly criticized the LDP's neoliberal policies over the past twenty years and stressed the need for "redistribution" by regulating the financial markets. That marks a clear contrast with the so-called "Abenomics" of the previous administration, whose commitment to trickle-down theory carefully avoided any talk of redistribution.

There is of course a lot of work still to do. I originally wrote this book with the goal of fusing degrowth and Marxist theory to update our vision of the post-capitalist world. In my home country, that made it something of a novelty. But there are already various movements in the West seeking to challenge the root causes of the current ecological crisis, and debates about degrowth and the climate crisis are much more current and robust in English-speaking countries than they are in Japan, even despite my best efforts. So I have a new hope for my book in its English version: that it help bring about new opportunities for collaboration and solidarity between East and West. We need to work together; our problems, after all, are global.

One way or another, the era of neoliberalism is over. Free markets, austerity, and small government cannot cope with the multi-stranded crises of capitalism, democracy, and ecology. In Japan, just as in the English-speaking world, we have tried them and are living with the consequences. Here, then, is an opportunity to open up a new political vision. My book does not offer a single definitive answer, but I hope that it will contribute to dialogue and social movements for the transition to a better, more just world. It is more important than ever to invent a clear and bright future. So let us work together. There is, in fact, no alternative.

—Kōhei Saitō, Tokyo, Japan, August 2023

SLOW DOWN

Introduction

Ecology Is the Opiate of the Masses!

What kinds of measures are you taking, personally, to prevent global warming? Have you bought a reusable shopping bag to reduce your reliance on disposable plastic ones? Do you carry a thermos so you don't end up buying drinks in plastic bottles? Did you buy an electric car?

Let me make one thing clear from the start: these good deeds are meaningless in the end. They can even cause more harm than good.

What do I mean by that? Simply that thinking such actions are effective countermeasures can prevent us from taking part in the larger actions truly necessary to combat climate change. They function like Catholic indulgences, allowing us to escape the pangs of our conscience about consumerism and look away from the real imminent danger around us, which in turn allows the forces of capital to swaddle our concerns in "environmental impact statements" and tuck them away beneath the form of deception known as greenwashing.

With this in mind, let us turn to SDGs—Sustainable Development Goals—as put forth by the United Nations and promoted by world governments and major industries. Do they have the power to change the overall global environment? The fact is, as you might have guessed by now, they won't work either. Many governments and major industries have conformed to various aspects of these SDGs already, yet these actions have proven unable to stop climate change. SDGs mainly function as an alibi, most effective at allowing us to avert our eyes from the danger right in front of us.

Long ago, Marx characterized religion as "the opiate of the masses" because he saw it as offering temporary relief from the painful reality brought about by capitalism. SDGs are none other than a contemporary version of the same "opiate."

The reality that we must face—that we must not flee from into the arms of a comforting opiate—is that we humans have changed the nature of the Earth in ways that are fundamental and irrevocable.

The effects of human economic activity have been so extensive that they led Paul Crutzen, Nobel Laureate in Chemistry, to declare that from a geological point of view, the Earth had entered a new era, one he dubbed the Anthropocene. He defined this as an era in which human economic activity has covered the surface of the Earth completely, leaving no part of it untouched.

Indeed, buildings, factories, roads, farmland, dams, and the like literally cover the Earth, and even the seas are awash in microplastics. Man-made materials are radically transforming the whole world. Among them, the material whose presence is most dramatically increasing due to human activity is atmospheric carbon dioxide.

As is well known, carbon dioxide is one of many greenhouse gasses. Greenhouse gasses absorb the heat given off by the Earth and radiate it back into the atmosphere. This greenhouse effect is what allows the Earth to maintain a livable temperature for living things, human beings included.

However, ever since the Industrial Revolution, humans have used more and more fossil fuels like coal and oil, releasing unprecedentedly enormous amounts of carbon dioxide into the atmosphere. Before the Industrial Revolution, the density of atmospheric carbon dioxide was around 280 parts per million (ppm), while by 2016, the level had passed 400 ppm even at the South Pole. This was the first time these levels had been reached in four million years. And they are being exceeded more and more every day, even as you read this.

Four million years ago, in the Pliocene Epoch, the average temperature of the Earth was warmer than it is now by around 35.6 to 37.4 degrees Farenheit, the ice shelves of Antarctica and Greenland were completely melted, and ocean levels were at minimum nineteen feet higher than today's. Research has shown that at times, they were as much as thirty-two to sixty-five feet higher.

Is climate change in the Anthropocene moving us toward the same conditions? Whether we reach that point or not, it seems clear that human civilization is facing a threat to its very existence.

The economic growth brought about by modernization promised us a richer lifestyle. But the ironic truth revealed by the environmental danger posed by the Anthropocene is that economic growth itself is what is destroying the very basis of what humans need to thrive.

The ultrarich living in the developed world may be able to maintain their heedlessly luxurious lifestyle even as climate change continues its rapid advance. But most of us ordinary people barely getting by each day will lose our way of life completely and be forced to scramble desperately just to survive. This hard truth should have become obvious to all during the COVID-19 pandemic.

There have been increasing calls during the pandemic for a fundamental rethinking of the way things have been done up until now, a way of doing things that has resulted in the dramatic widening of gaps between classes and the destruction of the global environment. The "Great Reset" proposed at the World Economic Forum at Davos in 2020 is a representative example of this. Even the superrich and global elites are recognizing a systemic transformation of our current economic system that increases economic inequality and environmental degradation.

But we must not entrust the salvation of the Earth's future to the emergency responses dreamed up by politicians, experts, and other elites. Leaving it to others in this way inevitably leads to the ultrarich prioritizing themselves only. For this reason, the best option for a better future is for ordinary citizens to step up as individuals, to testify to their experiences, to raise their voices and take action on their own initiative. Though it is not enough to scream into the void or act simply for the sake of "doing something"—if it doesn't go well, all this will accomplish is wasting even more precious time. It's thus essential to employ appropriate strategies and head in the right direction as we move forward.

To determine which direction is right, we must trace the current climate crisis to its root cause. The root cause is capitalism, and understanding this is key. The enormous increase in carbon dioxide emissions began, after all, with the Industrial Revolution—that is, when capitalism first began to truly operate in the world. It was shortly after this occurred that someone appeared who was able to clearly comprehend and analyze capitalism's nature. This was, of course, the German thinker Karl Marx.

This book proposes to analyze the entanglements of nature, society, and capital as they exist in the Anthropocene while making occasional references to Marx's *Capital* along the way. By no means do I wish to rehash Marxism as it presently exists. Rather, I intend to excavate and build upon a completely new, previously unexplored facet of Marx's thought that has been lying dormant for the past 150 years.

It is my hope that *Slow Down* will help to unleash the imaginative power necessary for us to create a better society in the age of climate crisis.

1

Climate Change and the Imperial Mode of Living

THE SINS OF THE NOBEL PRIZE IN ECONOMICS

In 2018, the Nobel Prize in Economics was awarded to Yale professor William Nordhaus, whose area of specialization is the economics of climate change. It might seem like fantastic news that such a person won the Nobel, a sign that modern society was finally beginning to confront the climate crisis directly. But some members of the environmental activist community raised their voices in sharp criticism of the decision instead.[1] Why? The main focus of their criticism was an article Nordhaus published in 1991. This article forms the basis of the line of research that led to his winning the Nobel.[2]

Speaking of 1991, this was right after the Cold War ended, on the eve of the wave of globalization that would go on to produce unprecedented emissions of carbon dioxide into the atmosphere. Nordhaus was among the earliest to incorporate the climate change problem into the field of economics. He did so by, as one might expect of an economist, proposing the introduction of a carbon tax and creating a model to determine an optimal rate of carbon dioxide emission reduction.

The problem lies with the optimal measures he proposes in his paper. To combat climate change, it is imperative that greenhouse gas emissions decrease. On the other hand, if emissions reduction goals are set too high, it might hinder economic growth. Therefore, he asserts, what we need is "balance." But in my view, Nordhaus's proposed "balance" leans much too far toward the side of economic growth.

According to Nordhaus, it is more beneficial to continue at the present rate of economic growth than worry excessively about climate change. Economic growth enriches the world, and this wealth will lead to the creation of new technologies. Therefore, if we allow economic growth to flourish, future generations will use the best technologies to combat climate change more effectively. In effect, Nordhaus is saying that because of the advancement of economic growth and new technologies, there is no need to preserve the natural world in its present state for future generations.

Given this premise, the optimal rate of carbon dioxide reduction Nordhaus proposes would result in a rise in average global temperatures by 38.3°F by the year 2100.[3] In other words, the optimal measures proposed from the point of view of economics will not combat climate change in any substantial way. In 2016, the Paris Agreement proposed the goal of limiting the rise in average global temperatures to no more than 35.6°F (and if possible, 34.7°F) higher than they were before the Industrial Revolution.

But even this goal of 35.6°F represents quite a dangerous change, and many scientists are sounding the alarm that we must keep the rise in temperatures below 34.7°F. And yet, Nordhaus's model would produce a rise of 38.3°F.

A rise in average global temperatures of 38.3°F or more would of course result in catastrophic damage, especially in the Global South, including African and Asian countries. However, the contributions such countries make to the global GDP is small. Agriculture would of course sustain serious damage. But agriculture makes up a "mere" 4 percent of the global GDP. A mere 4 percent isn't much, right? No matter if the damage extends to the people who live in these African and Asian countries. This is the line of thought propagated by the researcher who won the Nobel Prize in Economics.

Having won the Nobel Prize, Nordhaus's influence on the field of environmental economics is naturally great. Mainstream environmental economics emphasizes the limits of nature and the scarcity of resources. Calculating optimal distribution according to limits and scarcity is the special province of economics as a discipline. The goal is to create a "win-win" solution from this calculation that would benefit both nature and society.

This is why it's so easy to accept Nordhaus-style solutions. They are indisputably effective as strategies for economists to raise their profile among international organizations and so on. But the cost of this prominence is the justification by economists of lackadaisical climate change policies that are little better than doing nothing at all.

Nordhaus's style of thought naturally influenced the Paris Agreement as well. The goal set by the Paris Agreement for limiting the rise in global temperatures to 36.5°F amounts to little more than a form of lip service. Even if every country abided by the agreement, there are signs that global temperatures would rise by 37.94°F anyway.[4] It's only natural that world governments would be inclined to privilege economic growth and put off dealing with the problem at hand.

For this reason, it's no mystery why global rates of carbon dioxide emissions continue to rise every year even as the media is filled with buzzwords like SDGs (Sustainable Development Goals) and ESG (Environmental, Social, and Governance) as measures to encourage more sustainable and ethical business models and investments. There is no more time to waste in a world where average global temperatures have already risen by more than 33.98°F. Yet even so, the essence of the problem is treated as still undetermined, and the climate crisis of the Anthropocene continues to worsen.

POINT OF NO RETURN

Now, there's something I must clarify before moving on. The climate crisis is not something that will begin sometime around 2050. The crisis has already begun.

In fact, aberrant weather events known as one-in-a-hundred-year events have begun to occur every year in countries all over the world, a state of affairs that has come to be referred to as the "new normal." But this is actually only the beginning. The point of no return is approaching—the point at which a series of rapid changes will occur that will make it impossible to ever return to how things were.

For example, the temperature in Siberia reached 100.4°F in June 2020. This may be the highest temperature ever recorded in the Arctic Circle. If the permafrost there were to thaw, large amounts of methane gas would be released, which would accelerate climate change even more. On top of that, there's the risk of mercury leaking into the environment, as well as of bacteria and viruses (such as the anthrax virus) being released. Polar bears would also, of course, lose the last of their habitat.

The dangers compound as the crisis worsens. Further, once the climate change time bomb goes off, it will set off a chain reaction of crises like dominoes falling. This will lead to a level of destruction unable to be stopped by human hands.

Therefore, to prevent this catastrophe, scientists recommend that average global temperatures must not rise above 34.7°F more than pre–Industrial Revolution levels by 2100.

Temperatures have already risen to around 33.8°F higher than those levels, which means that we must act now to keep them below 34.7°F. Speaking concretely, this would entail lessening carbon dioxide emissions by nearly half by 2030 and reducing net emissions to zero by 2050.

If, on the other hand, emissions continue at their present rate, the rise in global temperatures will pass 34.7°F by 2030 and may even reach a maximum of 39.2°F by 2100. There are, of course, attempts to reduce greenhouse gas emissions occurring right now all over the world, including the Paris Agreement, but these attempts are woefully inadequate; it's said that they will still result in a rise of 37.76°F over preindustrial levels by the end of this century, a rise that is close to Nordhaus's model favoring technological advancement and economic growth as the most efficient way to combat climate change.

THE DAMAGE FORECAST FOR DEVELOPED COUNTRIES

The 34.7°F temperature rise has yet to occur, but already in the early 2020s, we've seen flooding in Pakistan that submerged a full third of the country's land while Africa faced massive starvation due to a severe drought. If temperatures continue their sharp rise, there is no reason to think that developed countries will emerge unscathed. A 35.6°F rise in temperature would spell the extinction of coral and deal a serious blow to the fishing industry. Heat waves in summer will have grave effects on harvests. Wildfires in dry regions like California and Australia will worsen, while every summer, typhoons and hurricanes will continue to grow and batter coastlines worldwide.

Torrential rainfall will also get worse. Where I live, in Japan, the cost of the damage caused by the 2014 torrential rains in western Japan exceeded one trillion, two hundred billion yen. Ever since that first disaster, Japan has seen comparable torrential rains every year all over the country, causing the cost of rain-related disasters to keep rising.

Moreover, the thawing of the ice sheets at the South Pole and similar places is predicted to lead, in eighty years' time, to devastating rises in ocean levels. If the worst-case scenario comes to pass, there is a possibility that American cities like New York and San Francisco, including many of their famous landmarks, will be submerged. San Francisco's world-renowned Fisherman's Wharf will be completely underwater, and One World Trade Center in Manhattan will be reachable only by boat.

At the global scale, the number of people who would have to evacuate their current homes is in the hundreds of millions. The global food supply would no longer be able to support the population. And economic damage would also be serious; some calculate the cost at upward of $27 trillion annually. This kind of damage would continue indefinitely.

THE GREAT ACCELERATION

Of course, most of the responsibility for climate change falls squarely on the shoulders of those of us living in the Global North. According to British charity Oxfam, the amount of carbon dioxide emitted by the richest 10 percent of the planet's population makes up more than half of total worldwide emissions. Cars, airplanes, large houses, meat, wine—sustaining lifestyles that include these things requires a huge amount of resources and energy to be wasted for the benefit of a very small portion of humanity. Considering the sheer scale of the irreversible effects that climate change will bring about for future generations, it's unforgivable for us not to do something about it while we still can. Now is the time to call for great change and to concretely bring it about. I propose that this great change should be nothing less than a challenge to the capitalist system itself.

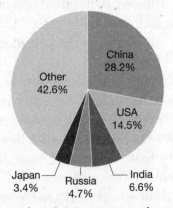

Figure 1. Share of total CO2 emissions by country (2017)
Based on "The Institute of Energy Economics," ed. *Energy Data and Modeling Center Energy and Economics Statistical Summary, 2020* (Energy Conservation Center, Japan, 2020)

But before presenting this seemingly unrealistic demand, I must first address the origins of the environmental crisis now presenting itself in the form of climate change.

Here I would like to include the results of studies conducted by Will Steffen and other researchers belonging to the Australian National University Institute for Climate Energy & Disaster Solutions. According to them, human economic activity following the Industrial Revolution greatly increased the burden on the environment. As the human population and energy consumption increased, the atmosphere became filled with more and more carbon dioxide while the destruction of rainforests reached devastating levels. This activity increased even more rapidly after the end of World War II, as did the resulting burden on the environment, and this dramatic increase is referred to as the Great Acceleration. This acceleration only increased following the end of the Cold War. This age of acceleration cannot continue indefinitely. The trajectory of the Anthropocene is heading nowhere but toward total destruction.[5]

But how did this state of affairs come about? To clarify the origins of the crisis, we must first understand the relationship between the globalization of capitalism and environmental destruction. This will be the subject of the rest of this chapter.

THE REPEATED MAN-MADE DISASTERS INFLICTED ON THE GLOBAL SOUTH

Looking back at the history of capitalism, we can see various sorts of tragedies underpinning the enriched lifestyles enjoyed by those in the Global North. One might say that the contradictions of capitalism are distilled into the phenomenon of the Global South.

Even if we concentrate on just the most spectacular disasters of recent years—oil spills in Russia, Mauritius, and Venezuela, the wildfires devastating the Amazon rainforest precipitated by the rapacious spread of multinational agribusiness, two failed dams in central Michigan caused by excess rainfall—there are too many to count.

The scope of the damage is wide. The 2019 collapse of a dam in Brumadinho, Brazil, resulted in the deaths of more than 250 people. The dam was owned by Vale, one of the top three resource exploitation companies in the world, and was used to dam up iron ore tailings—the slurry of water and mineral by-products produced by the ore sorting process.

A similar accident had occurred previously at another Vale tailings dam in 2015, but this time, careless mismanagement resulted in a collapse that caused hundreds of thousands of tons of toxic slurry to engulf a nearby village in its flow.[6] Tailings covered the entire area, resulting in the pollution of rivers and grave damage to the ecosystem overall.

Are these kinds of disasters simply the results of bad luck? Of course not. The danger that such accidents would occur was repeatedly pointed out by experts, workers, and the people living in the area. And yet, government and industry prioritized cost cutting over developing effective measures to prevent the disaster.[7] These are entirely predictable disasters known as man-made disasters.

Even so, the Global North may well choose not to care about these incidents occurring in distant places like Mexico and Brazil. There may be readers who think these events have nothing to do with them. But make no mistake, all of us living in the developed world are complicit in these man-made disasters. Furthermore, these incidents can happen in the Global North, too; see, for example, the Keystone Pipeline oil spill in 2022 and the train derailment in East Palestine, Ohio, in 2023. These are classic cases of the capitalist push for profit at all costs leading to climate catastrophe and lasting consequences for the health of local populations.

In any case, the iron in our cars, the gasoline they consume, the cotton woven into our clothes, the beef in our *gyūdon* bowls: these are the things that come to us from these "distant" places. Our rich lifestyles would be impossible without the plundered natural resources and exploited labor power of the Global South.

THE IMPERIAL MODE OF LIVING IS BASED ON THE SACRIFICE OF OTHERS

German sociologists Ulrich Brand and Markus Wissen gave a name to the lifestyle of people in the Global North that is based on the extraction of energy and natural resources from the Global South. They call it the Imperial Mode of Living (*imperiale Lebensweise*).

The Imperial Mode of Living refers, essentially, to the societies of the Global North that rely on large-scale production and consumption. This is what makes our rich lifestyles possible. Beneath this surface, there exists a structure by which the cost of our consumption is extracted from the lands and labor of the people of the Global South. Without the exploitation of others who pay the cost, the Imperial Mode of Living would be unsustainable. Lowering the standard of living for those in the Global South is a prerequisite for the workings of capitalism, and the power imbalance between North and South is no anomaly—it is, in fact, a result of the system functioning normally.[8]

We experience this way of life as desirable, though, and are loath to give it up. If we were to acknowledge the state of things in the Global South, we would be forced to lower our own standard of living. Our way of life is, in fact, a terrible thing. We are all complicit in the Imperial Mode of Living.

Let me give an example of what I mean. The fast fashion we have so thoroughly incorporated into our way of life is often produced by Bangladeshi laborers working in the worst conditions. You may recall the famous incident of 2013 at the Rana Plaza in Dhaka, where a huge building that housed five garment factories collapsed, resulting in more than a thousand people losing their lives.

Further, the cotton that goes into the clothing produced in Bangladesh is cultivated by poor farmers in India working in oppressive 104°F+ heat.[9] Ever-increasing demand from the fashion industry has led to the widespread use of genetically modified cotton plants. As a result, farmers lose possession of their own seeds and are forced instead to borrow money to purchase genetically modified ones, along with chemical fertilizer and pesticides, every year. When crops fail due to drought or heat waves, the farmers end up accumulating more and more debt, and it's not uncommon for this to drive them to suicide.

The tragedy I've outlined here depends on the normal functioning of global capitalism, just as, for structural reasons, the Global South depends on the production and consumption that goes into the Imperial Mode of Living.

As I've previously stated, Brazilians understood the danger of the tailings dam in Brumadinho. The exact same type of accident had happened before. And yet, despite this, they were forced to continue mining; the laborers had to work at the mining site to support their own lifestyles, and they were also forced to live close to the mine.

In the case of the garment factories in Rana Plaza, the workers had noticed irregularities in the walls and pillars, but their warnings were ignored. Moreover, the farmers in India realize that pesticides harm their bodies and the natural world. But they are forced to keep working and producing regardless, in order to satisfy worldwide demand, as the fashion industry and agribusiness keep growing and growing to satiate the unlimited desires of consumers who want whole new wardrobes every season.

In this way, increases in human and environmental sacrifice result in increases in profit for major industry. This is the logic of capital.

EXTERNALIZATION SOCIETY RENDERS SACRIFICE INVISIBLE

Of course, the hard-to-hear truth outlined above has been pointed out many times before. But as soon as we throw a little money at it via charitable donations, we forget about it again. This forgetting is possible because the incidents in question are rendered invisible in our daily lives.

The Munich University sociologist Stephan Lessenich has pointed out that this passing along of costs to somewhere far away, and rendering them invisible because of the twenty-four-hour news cycle and the attention economy, is indispensable to the "richness" enjoyed by societies in the Global North. He critiques this tendency, calling it "externalization society."

Those in the Global North enjoy rich lifestyles enabled by the sacrifices of those in the Global South. Further, Lessenich points out that those in the Global North work to maintain this exceptional status not just for today, but for the foreseeable future. "Externalization society" refers to the process by which a society tirelessly creates an "outside" so as to pass along the various burdens necessary to maintain itself. This is the only way our present society has been able to thrive and flourish.[10]

BOTH WORKERS AND THE ENVIRONMENT ARE OBJECTS OF EXPLOITATION

The relationship between capitalism as it is practiced in Global North countries and the sacrifices of the Global South can perhaps be summed up by referring to the "world systems" approach of American sociologist Immanuel Wallerstein.

As Wallerstein saw it, capitalism relies on an opposition between "core" and "periphery." Cheap labor is extracted from the periphery known as the Global South, and the core raises its profits by driving down the price of the goods produced by that labor. This unequal exchange of labor power is what brings about the overdevelopment of developed countries and the underdevelopment of developing ones, according to Wallerstein.

However, as capitalism's global reach has extended to every corner of the world, new frontiers to plunder in this way have disappeared. (The digital space is today's latest and last frontier of capitalism.) The profit-seeking process has reached its physical limit. The resulting plunge in profits makes economic growth and the accumulation of capital difficult, leading some to even declare that this spells the end of capitalism.[11]

But what I want to point out in this chapter is what comes next. Human labor power is the object of exploitation that most concerns Wallerstein, but this reveals only one side of how capitalism works.

The other side of things is, naturally, the global environment. The object of capitalism's exploitation is not just the labor power of the periphery but also the environment of the entire Earth. Natural resources, energy, and food are all plundered from the Global South via unequal exchange with developed countries. Capitalism uses humans as tools for accumulating capital but can profit from the natural world by simply plundering its resources directly. This is one of this book's most fundamental assertions.[12]

Therefore, it is only to be expected that as long as it aims for unlimited economic growth, a global system of this sort will place the world's environment in danger.

THE EXTERNALIZATION OF
ENVIRONMENTAL BURDEN

Simply put, if we expand Wallerstein's theory, we can see how the economic growth of the core has necessitated the plundering of natural resources from the periphery while at the same time shifting the costs underlying this growth onto the periphery as well.

Let's look at the example of palm oil, which has played a major role in contributing to the shadow our food consumption has cast across the world. Not only is palm oil cheap, it doesn't oxidize easily, and this has led to its widespread use in processed food, sweets, and fast food. When eating out, it can be difficult to find foods without palm oil.

Palm oil is produced in Indonesia and Malaysia. The area needed to cultivate the palms used to produce palm oil has increased exponentially since the beginning of this century, leading to rapid deforestation as the tropical rainforest is cleared to make room.

This sudden spike in palm oil production affects more than just the ecosystem of the tropical rainforest. This kind of large-scale development has destructive effects on people who depend on the rainforest to sustain their way of life. For example, the clearing of the rainforest to make way for farmland has caused soil to erode, fertilizer and agrichemicals to pollute rivers and streams, and fish populations to decrease. People living in the region used to depend on the fish in the streams for protein, and without that, they're forced to spend more money on processed food. This has led to the people in the area sullying their hands with the illegal trade of species on the brink of extinction, such as tigers and orangutans, in order to obtain that money.

In this way, the inexpensive, convenient lifestyle of the core is underwritten not only by the exploitation of the periphery's labor but also by the extraction and destruction of natural resources and the environmental burden that goes with that. Furthermore, the damage brought about by the environmental crisis is not borne equally by the global population. The environmental burden tied to the production and consumption of food, energy, and other raw materials falls on the shoulders of some much more than others.

As Lessenich, the vocal critic of the Global North as externalization society, puts it, the primary condition making our rich lifestyles possible is the passing of these burdens onto the people and natural resources of "some faraway place" so that the true cost is never paid by the end consumer.

THE DENIAL OF WRONGDOING AND PROCRASTINATION'S JUST DESERTS

The Imperial Mode of Living is reproduced again and again as we go about our daily lives while the violence of sustaining it occurs in some distant place and is, as such, rendered invisible.

We hear the words "environmental crisis" and, like a Catholic buying an indulgence, we purchase reusable shopping bags or organic cotton T-shirts. But there are always newer versions of these eco-bags and organic T-shirts hitting the market. Inspired by advertisements, we go out and buy the newest ones, and then the newest ones after that. And the sense of satisfaction and accomplishment we gain by buying this form of indulgence means we are unconcerned with the violence against the people and environments of the distant lands involved in producing a reusable bag or T-shirt. This is what it means to be caught up in greenwashing as devised by capital.

Those of us living in the Global North are not just compelled into ignorance of the transfer of our lifestyle's cost to the Global South. We are also compelled to internalize to an extreme extent the sheer desirability of the Imperial Mode of Living. We hold on to blissful ignorance and are afraid to look directly at the truth. "I don't know" evolves naturally into "I don't want to know."

But don't we know deep down that our comfortable lives come at the expense of others forced to live in comparative misery? We merely see this injustice as something that has nothing to do with us, and so we look away from it. We cannot stand to look the truth in the face, so although we know we are the cause of injustice, we secretly want the current order to continue.[13]

This is how the Imperial Mode of Living becomes ever more deeply entrenched, and our response to the current crisis ends up being put off to a future time that never comes. In this way, each and every one of us is complicit in perpetuating injustice. But the just deserts for this procrastination have already begun filtering back into the core.

ARE DEVELOPED COUNTRIES KINDER TO THE ENVIRONMENT? THE NETHERLANDS FALLACY

Once again, pointing these things out is nothing new. A similar debate erupted in the 1970s and '80s as issues of responsibility for environmental damage and the North-South problem came to the fore. One example of this is the so-called Netherlands Fallacy.

The way of life enjoyed by people who live in a country like the Netherlands places a heavy burden on the Earth. However, the air and water pollution in such a country is comparatively low. By contrast, countries in the Global South suffer many environmental issues, such as air and water pollution, waste disposal problems, and so on, despite the people in these countries living comparatively modest lifestyles.

What produces this seemingly contradictory state of affairs? One way to explain it is to point to technological advances. The technology brought about by economic growth enables "advanced" countries to reduce and dispose of pollutants that damage the environment.

But to celebrate the economic growth of developed countries for lessening the pollution in those countries is the essence of the fallacy here, which Nordhaus has fallen into, as seen at the beginning of this chapter. The environmental improvements occurring in developed countries result not just from technological advances but from passing along most of the negative by-products of economic development—resource extraction, waste disposal, and the like—outward onto the periphery represented by the Global South.[14]

To ignore the international transfer of the burden of environmental impact and to assume that the Global North has solved its environmental problems simply through technological advances and economic growth is what is known as the Netherlands Fallacy.

THE ANTHROPOCENE AND THE EXHAUSTION OF THE PERIPHERY

One way to understand the Anthropocene, when human economic activity has reached every corner of the world, is as an era when the periphery needed to perform this plunder-and-cost transfer has effectively disappeared.

Capital has always strived to extract everything it can—all the oil, all the nutrients from the soil, all the rare metals, and so on—from the world. This is known as extractivism, and it has placed an enormous burden on the Earth. But just as the frontiers needed to gain profits from cheap labor power have started to disappear, the cheaply available natural resources of the periphery have also begun to reach the point of exhaustion, no longer able to perform their role as repositories of displaced environmental damage and plunderable riches.

No matter how well capitalism seems to be functioning, there will always be a limit on the resources available for exploitation. The disappearance of a sufficient "outside" to externalize into a periphery has resulted in the negative consequences of extractivism's spread coming back home to roost in the Global North.

This is a problem that even the power of capital cannot overcome. Capitalists strive toward unlimited increases in value, but the Earth is not unlimited. Once the periphery is exhausted, things can no longer function the way they have up until now. A crisis will begin. This is the essence of the crisis plaguing the Anthropocene.

Could there be a better example of this than the accelerating climate change crisis? As the periphery is exhausted, the disasters associated with climate change—the superhurricanes, the wildfires, and so on—become more and more conspicuous even in the developed world.

So the question becomes, now that time has definitively run out for addressing climate change, what can we possibly do to stop it?

WASTING TIME AFTER THE END OF THE COLD WAR

There's a famous quotation attributed to the economist Kenneth E. Boulding that goes, "Anyone who believes that exponential growth can go on forever in a finite world is either a madman or an economist." Half a century has passed since he said that, and yet we still single-mindedly pursue economic growth despite how severe the climate crisis has become or how much destruction we visit upon the Earth. This is how deeply this economist-minded way of thinking has rooted itself into our daily lives. We may all be madmen these days.

Children, though, remain sane. It took young Swedish environmental activist Greta Thunberg to expose the hypocrisy of the measures to combat climate change dreamed up by adults. Only fifteen years old at the time, this high schooler, who became famous for her School Strike for Climate sharply criticized politicians who, in an attempt to court votes, "only speak of green eternal economic growth." She said this during an address to the United Nations Climate Change Conference (COP24) in 2018.[15]

Thunberg's point was that as long as capitalism prioritizes economic growth above all else, the climate change problem will never be solved. And indeed, it's hard not to agree with this assessment. After all, capitalist society wasted its opportunity to combat climate change during the crucial thirty years following the end of the Cold War, preferring instead to concentrate on pursuing the money-making opportunities opened up by globalization and financial market deregulation.

Looking back to 1988, NASA scientist James Hansen testified before the US Congress that he was "99 percent confident" that climate change was caused by human activity. That year also saw the formation of the Intergovernmental Panel on Climate Change (IPCC) by the United Nations Environmental Programme (UNEP) and the World Meteorological Organization (WMO).

There was hope for climate change to be solved with international agreements and treaties. And indeed, if solutions had truly started to be implemented at that time and carbon dioxide emissions had started decreasing even at the leisurely pace of 3 percent per year, it would have been possible to solve the problem this way.

But Hansen's warning turned out to be ill-timed. The Berlin Wall fell soon after he spoke, ushering in the fall of the Soviet Union, and American-style neoliberalism spread throughout the entire world. Capitalism was handed a new frontier to exploit in the form of the cheap labor and marketplaces now accessible in the former Soviet Union and its satellite states.

This great expansion of economic activity resulted in a similarly great acceleration in the consumption of resources. To take one example, almost half of humanity's total consumption of fossil fuels throughout history occurred in the years following the end of the Cold War in 1989.[16]

Nordhaus's article on climate change, with its naïve predictions about how much carbon dioxide emissions needed to be reduced in order to combat it, came out around that time as well. This is how a crucial thirty years that could have been used to develop effective climate change solutions was wasted, resulting in the greatly worsened situation we find ourselves in now.

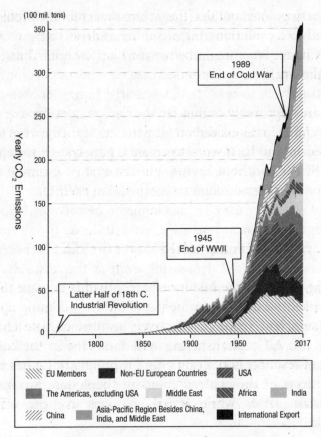

(100 mil. tons)

Yearly CO$_2$ Emissions

1989
End of Cold War

1945
End of WWII

Latter Half of 18th C.
Industrial Revolution

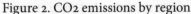

EU Members	Non-EU European Countries	USA
The Americas, excluding USA	Middle East	Africa	India
China	Asia-Pacific Region Besides China, India, and Middle East	International Export

Figure 2. CO2 emissions by region
Based on data from the Carbon Dioxide Information Analysis Center
(CDIAC) and the Global Carbon Project (GCP)

This is why Greta Thunberg's criticism is so forceful and passionate—it comes from her anger at the irresponsibility of adults who wasted this precious chance to do something because they could only think about what was right in front of their faces. The actions of politicians and other elites who still prioritize economic growth above all else and whose attitudes haven't changed one bit only pour fuel on the fire of this anger. As Thunberg put it, "You don't listen to the science because you are only interested in solutions that will enable you to carry on like before. Like now. And those answers don't exist anymore. Because you did not act in time."[17]

Thunberg pointed out that the system that created a problem will never lead to its solution and ended her address to the COP24 by saying, "Change is coming, whether you like it or not." Young people around the world have responded with fervent support for her and her message.

If we are to properly respond to the clarion call of young people, we adults must first look directly at what lies at the heart of the present problem and then work to create a new system to replace it. Perhaps it goes without saying, but the system Greta Thunberg indicts for offering no solutions is capitalism itself.

MARX'S CLIMATE CRISIS PROPHECY

A passing glance at the history of capitalism is enough to tell us that the likelihood of government and industry coming up with a plan of sufficient scope to effectively combat climate change is basically nil. All capitalism has ever done is shift the costs and burdens of resource extraction onto the peripheries it has created. The ill effects of the contradictions in the system are displaced onto distant lands while actual solutions are put off again and again.

The fact is, capitalism's creation of peripheries onto which to shift the costs of its workings—and the injustice of this—was pointed out as far back as the mid-nineteenth century by Karl Marx himself.

Marx's main point was this: capitalism displaces its contradictions elsewhere and thus renders them invisible. This displacement only makes the contradictions worsen, deepening the quagmire into an increasingly disastrous state of affairs. Capital's attempts to displace its ill effects can only result in collapse. According to Marx, this is the ultimate, insurmountable limit of capital.

To ascertain capitalism's limits, we should use Marx's writings to trace three fundamental forms of displacement or "shift": technological, spatial, and temporal.

TECHNOLOGICAL SHIFT—DISTURBING
THE ECOSYSTEM

The first form of displacement is the attempt to overcome an environmental crisis through technological advancement, as Nordhaus suggested. Marx's example of this is soil depletion due to agriculture. His main reference was the work of his contemporary, the German chemist Justus von Liebig, who criticized then-current agricultural practices as "robbery of the soil."

According to Liebig, nutrients in the soil—namely inorganic substances like phosphorous and potassium—are converted by the natural breakdown of rocks into forms usable by plants. Because this process is extremely slow, the amount of nutrients in the soil that can support plant growth at any given time is finite. Therefore, we must use fertilizers to replace the inorganic substances taken out of the soil by crops like grains in order to maintain the soil's fertility. Liebig called this the law of compensation. In short, this law states that agriculture can only be sustained if the nutrients in the soil are sufficiently replaced.

However, the advancement of capitalism led to a division of labor between farming villages and urban centers whereby the crops cultivated in rural locations were sold to workers living in cities. This meant that the nutrients extracted by the crops consumed in the city never returned to the soil from whence they came. Instead, they were consumed and digested by urban workers and then deposited into flush toilets that expelled them into rivers and streams.

Problems arose from conducting agriculture under capitalism as well. Agricultural businessmen became concerned primarily with the short-term bottom line, preferring to profit from serial cultivation of the same land over leaving fields fallow to allow their nutrients to be renewed. Funds used to maintain the soil, such as those used for irrigation systems and the like, were also cut to the bare minimum. Capitalism always prioritizes short-term profits. This produces a "rift" in the soil's metabolic cycle whereby nutrients are never returned to the land and are instead only extracted from it, leading to its exhaustion. Liebig raised the alarm about the danger this represented, calling the irrational agricultural management system whereby long-term sustainability is sacrificed in the name of short-term profitability a "robbery system" that would eventually lead to the collapse of European civilization.[18]

However, history shows that the depletion of soil did not lead to the civilizational crisis Liebig foretold. Why not? One answer is the discovery of the Haber-Bosch process by which ammonia can be created industrially, a discovery that allowed large amounts of chemical fertilizer to be made at a low price. But this discovery did not in fact heal the rift produced by the contradictions of capital. It merely *displaced* its effects.

The Haber-Bosch process produces ammonia (NH_3) not just by using nitrogen (N) from the air but also hydrogen (H) extracted from fossil fuels (primarily natural gas). This of course means that a huge amount of fossil fuels are needed to satisfy the fertilizing needs of the world's farmland. Indeed, the amount of natural gas used to create ammonia this way constitutes 3–5 percent of the world's total output.[19] In other words, modern agriculture replaces the original nutrients in the soil by using up another nonrenewable natural resource. And of course, the production process also emits large amounts of carbon dioxide into the atmosphere. This is the fundamental nature of the contradictions produced by technological shift.

Furthermore, the development of industrialized agriculture that uses large amounts of chemical fertilizer has led to myriad other issues caused by nitrogen compounds polluting the environment, from nitric acid contamination of groundwater to the red tide blooms caused by the eutrophication of aquatic ecosystems. In this way, technological shift eventually triggers large-scale environmental problems ranging well beyond the simple exhaustion of a single tract of land.

But even this is not the end of the story. Soil ecosystems disturbed by large-scale use of chemical fertilizers end up losing their ability to retain moisture, while crops and livestock become increasingly vulnerable to disease. Yet the marketplace demanding cheap agricultural products is unyieldingly uniform in scale and size, necessitating modern agriculture to use more and more chemical fertilizer, agrichemicals, and antibiotics to meet this growing demand. And it goes without saying that these chemicals also end up leaking into the environment and disturbing ecosystems.

Even when damage occurs, the industries at the root of these problems claim that no cause-and-effect relationships can be proven, thus avoiding offering any compensation to those affected. And of course, many environmental problems are of a sort that even if compensation were forthcoming, things could never be restored to their original state. Technological shift does not solve problems. Rather, the excessive use of technological solutions only deepens the contradictions underlying those problems.

SPATIAL SHIFT—EXTERNALIZATION AND ECOLOGICAL IMPERIALISM

After technological shift, the next form of displacement I want to address is spatial. Marx addresses this form of displacement using the example of soil depletion as well.

The Haber-Bosch process had yet to be discovered in Marx's time, so the main form of fertilizer used to replenish the soil was guano. Huge populations of seabirds gather off the coast of Peru and deposit enormous amounts of excrement there, creating a petrified concentration of sediments referred to as guano. These deposits build up into veritable islands of the stuff.

Since guano is the dried excrement of birds, it contains many inorganic substances necessary to cultivate crops, and gathering and using it is relatively simple. In fact, Indigenous people in the area traditionally used guano to fertilize their own crops. The European who discovered the marvelous utility of guano was Alexander von

Humboldt, who conducted a biogeographical survey of South America at the beginning of the nineteenth century.

Following its discovery, guano became famous as the force that would save the world from soil depletion, and vast amounts of it were exported from South America to Europe and the United States. Thanks to guano, English and American farmland could be sustained and continue to produce the foodstuffs necessary for the urban workforce.

But this, too, ultimately failed to heal the rift. Large numbers of workers were pressed into service to extract guano and send it far away. This led to the violent exploitation of both the Indigenous peoples of South America and as many as 90,000 Chinese laborers. Moreover, the sudden decline in seabird populations in the area led to a similarly sudden decline in the resource guano that represented.[20] The exhaustion of this resource ended up touching off the Guano and Saltpeter Wars (1864–66 and 1879–84, respectively).

As this example makes clear, solving a problem through displacements that benefit only the core constitutes what we now call ecological imperialism. Ecological imperialism relies on the plunder of the periphery while shifting the ill effects of the problems brought about by that plunder back onto the periphery as well. In this way, these actions end up placing even heavier burdens on the local peoples and ecosystems that struggle to sustain themselves, further deepening the severity of the contradictions they were meant to resolve.[21]

TEMPORAL SHIFT—*APRÈS MOI, LE DÉLUGE!*

The last of the three forms of displacement is temporal shift. Marx's example of this form of displacement was the overharvesting of forestland, but the clearest example of how it works in the present age is climate change.

There is no question that large-scale consumption of fossil fuels causes climate change. At the same time, the effects of climate change do not all appear immediately. There is sometimes a time lag of decades before you can witness these effects. Capital uses this lag to reap as much profit as possible from the drilling equipment and pipelines it has already invested the money to create.

Capitalism reflects the opinions of shareholders and business owners living in the present and therefore ignores the voices of future generations, creating yet another type of externality by shifting the burden of environmental damage to the future. The present generation profits from the sacrifice of future generations.

There may be those who see this form of temporal shift in a positive light, who think that in fact provides the time necessary for technology to advance to the point at which the crisis may be averted. And indeed, there are scholars who, like William Nordhaus, believe that instead of cutting carbon dioxide emissions so severely that it would have a negative impact on the economy, it would be wiser to encourage economic growth and allow societies to become rich enough that technological progress will take care of the problem for us.

But even if such marvelous new technologies were to be discovered, it would still take a long time for them to be adopted effectively across society. This wastes precious time, as the crisis will continue to worsen as the positive feedback loop speeds up while we wait for a hypothetical new technology to spread, resulting in a much more serious environmental crisis in the meantime. This might well mean that it will become too serious for this new technology to adequately address anymore. The promise of a technological solution will thus end in betrayal.

As the positive feedback effect is allowed to amplify, it will naturally begin to have a negative effect on economic activity as well. If new technologies are unable to catch up with the speed of the environment's destruction, it will soon be something humanity will be unable to address at all, and future generations will be doomed. And this will also, of course, negatively affect economic activity. In other words, future generations will not only be consigned to live in extremely cruel conditions, but their economic situation will also be dire.

This is perhaps the worst possible outcome of all, one that constitutes the most compelling reason not to leave climate change to be solved by technology but rather to stop it here and now by addressing its root cause.

THE PERIPHERY'S DOUBLE BURDEN

We have now explored the three displacements outlined by Marx. Capital will forever find ways to displace the negative effects of its workings onto the periphery it creates.

As a result, the periphery faces a double burden. After suffering from the plunder of ecological imperialism, it finds itself forced to bear an unequal burden of the destructive effects of capital's displacements.

For example, the South American nation of Chile cultivates avocados for export. They are highly prized by the Global North as part of a "healthy lifestyle"—that is, as part of the Imperial Mode of Living. The cultivation of avocados, affectionately called "butter of the forest" in Japan, necessitates the use of huge amounts of water. Further, because avocados take so many nutrients from the soil as they grow, once land is used to cultivate them, it's very difficult to use it to grow anything else. In other words, Chile has sacrificed its own water supply and its capacity to produce its own food to cultivate this export crop.

Chile is now facing major droughts as well as serious water shortages. These problems are attributable to climate change. And as we've already seen, climate change itself can be seen as the consequence of capitalistic displacement. In the midst of this dire situation, the COVID-19 pandemic occurred. But instead of their precious water being used for the handwashing and hygiene necessary to combat the pandemic, it was used to grow more avocados for export. This was due to the privatization of the water supply.[22]

In this way, the negative impacts of climate change caused by the consumerist lifestyle enjoyed by the developed world and the deleterious effects of the COVID-19 pandemic were visited most severely upon the area most vulnerable to both: the periphery.

THE WORLD WILL END BEFORE
CAPITALISM DOES

One way to see the situation is as one in which risk and chance are distributed extremely unequally. The periphery must always lose so that the core may win.

Of course, the core is not entirely exempt from the impacts of the natural world's worsening condition. But thanks to displacement, capitalism is unlikely to be dealt a blow that would seriously threaten its ability to function. To put it another way, by the time people in developed countries are ready to look the problem in the face, it will already be too late to save the ecology of this limited planet. The Earth will become uninhabitable for humankind before capitalism collapses.

The famous American environmental activist Bill McKibben puts it this way: "The diminished availability of fossil fuel is not the only limit we face. In fact, it's not even the most important. Even before we run out of oil, we're running out of planet."[23]

This quote still holds true if we replace the words "fossil fuel" and "oil" with the word "capitalism." We must remember that if the planet fails, it's game over for humanity. There is no planet B.

THE CRISIS MADE VISIBLE

As long as we look at things only from a short-term, surface perspective, capitalist society still appears to be functioning smoothly (even if this apparent smoothness is increasingly threatened by factors like war, pandemic, inflation, and the like). As countries like China and Brazil, which have historically acted as the repositories of displacement for the developed world, increasingly devote themselves to their own high-speed economic development, the available space on the planet to use as the convenient "outside" for externalization and displacement shrinks into almost nothing. It's logically impossible for all these countries to externalize the contradictions of their development at the same time. But the lack of an "outside" constitutes a death blow for our externalization society.

The reality is, as frontiers from which to extract cheap labor power disappear, profit rates drop, and the exploitation of workers within Global North countries intensifies. At the same time, as the externalization and displacement of environmental burdens onto the Global South reach their limit, the contradictions of capitalism will start to appear more and more within those countries. Areas within Global North countries will turn into the Global South. Those of us living in the Global North will begin to feel the worsening of labor conditions directly. And it will only be a matter of time before we suffer directly from the consequences of environmental problems like climate change as well. They will no longer be someone else's problems.

Looking back at Wallerstein's argument, we can see how these problems are interconnected; after all, we have only one Earth to call home. As externalization and displacement become more difficult, the tab we've been running up will ultimately have to be paid. We've been dumping plastic waste into the ocean and pretending it disappears, but now it's popping back up in our daily lives in the form of microplastics in our seafood and our water supply. The carbon dioxide we've emitted into the atmosphere for so long has triggered the climate change now causing the heat waves, hurricanes, and wildfires we find ourselves battling every year.

Furthermore, the influx of refugees from Syria has become a serious social problem for Europe, allowing for the rise of right-wing populism, threatening democracy there. And in fact, one of the causes of the Syrian civil war is said to be climate change. The famine brought about by the long drought afflicting the country forced people into desperation, paving the way for the outbreak of increasingly intense societal conflicts.[24]

The United States is in a similar position. Not only has it been suffering from an onslaught of increasingly large-scale hurricanes, but its southern border is overrun by caravans of refugees from Honduras. These refugees are fleeing not only violence and political instability but also bad harvests due to climate change and the desperate circumstances that go along with them.[25] In 2019, then-president Trump responded with cruelty to these climate refugees, detaining them in miserable conditions while refusing to allow them into the country and calling instead for a border wall. Many of these policies are still in place under President Biden.

The European Union, too, is pushing out refugees seeking safety, in this case through Turkey. But this sort of literal displacement can't continue forever. Climate change and climate refugees render the previously invisible contradictions of the developed world's Imperial Mode of Living visible as concrete objects and bodies and threaten to overturn the prevailing order once and for all.

THE GREAT DIVERGENCE

In this way, the exhaustion of the periphery is making it harder and harder to turn a blind eye to the impending crisis. We can no longer heedlessly proclaim, *"Après moi, le déluge!"* and ignore what happens to the world after we're gone. After all, *le déluge* is already lapping at our feet. Climate change is forcing humanity into nothing less than an overdue reckoning with harsh reality and a radical rethinking of the Imperial Mode of Living that has made us so dependent on extractivism and externalization.

As the pending impossibility of displacement becomes clearer, though, distrust and unease grow, leading to the strengthening of nationalist movements. Right-wing populists are using climate change to promote their cause. Food shortages, energy crises, immigration problems, and so on only serve to incite chauvinistic nationalism. The subsequent division of society along these lines worsens the danger to democracy. As a result, authoritarian leaders take power, bringing about a form of rule that might best be called "climate fascism."

But even in this moment of crisis, there is opportunity. Climate change can force those of us living in the Global North to finally face the consequences of our behavior. The exhaustion of the periphery makes us into victims as well. The hope is that this will lead to broader support for movements calling for a more just world and a radical transformation of our lifestyle.

To borrow the words of Wallerstein, this is the true bifurcation, a moment of systemic crisis, brought about by capitalism itself, leading to systemic chaos and the emergence of new anti-systemic movements. The exhaustion of the periphery is bringing us all closer to a historic turning point that will render the present system unable to function. Right before he passed away, Wallerstein said, "The past 'normality' of externalization is a distant memory."[26] If externalization can no longer occur, neither can the accumulation of capital—at least not how it has occurred up until now. The environmental crisis will also continue to worsen. As a result, the justification for the continuation of the capitalist system will be undercut, and resistance movements opposing that system will gain strength and magnitude.

This is why Wallerstein left us with the thought that the present moment—a time when the periphery has reached the point of exhaustion—is a turning point in history. As capitalism collapses, will the world be plunged into chaos, or will a different, more stable social system replace it? This new "bifurcation" brought about by the end of capitalism has in fact already begun.[27]

Rosa Luxemburg's famous slogan, "Socialism or barbarism," feels all the more relevant again as we approach the Great Divergence of the twenty-first century. But how can we avoid barbarism? What seems sure is that it is already far too late for an incremental approach to reform.

What sort of large-scale, thoroughgoing measures must we take, then? In the next chapter, I will examine one such attempt, the "hope" offered by the Green New Deal proposed in Europe and the United States.

2

The Limits of Green Keynesianism

THE GREEN NEW DEAL—A NEW HOPE?

In the previous chapter, we saw that capitalism is a system that exploits not just humankind but the environment as well. Further, it fuels economic growth by shifting the cost of that development onto the periphery. As long as this externalization of costs runs smoothly, those of us living in the Global North can enjoy rich lifestyles and avoid suffering the consequences of environmental crises. This is how we've been able to avoid thinking seriously about the true cost of our expansive lifestyle.

The capitalist system itself is the main reason why the environmental crisis has become as serious as it has. Coddled by the invisibility of our lifestyle's costs, those of us living in the Global North have been able to turn our backs on reality, ignoring the faint inkling of awareness tickling the edges of our consciousness, and now we've run out of time to take measures to address the danger.

As we've procrastinated, the true cost of our lifestyle has become steadily more difficult to ignore. As the time left before we pass the point of no return shrinks to almost nothing, the possibility of implementing major unprecedented policies at the governmental level is being debated around the world.

One plan that has inspired much hope and expectation is the Green New Deal. Prominent pundits in the United States like Paul Krugman, Thomas Friedman, and Jeremy Rifkin have called for its adoption, stressing its necessity. Politicians around the world, including Bernie Sanders, Alexandria Ocasio-Cortez, and Yanis

Varoufakis, have run for office with the Green New Deal as part of their platforms, albeit each with different details.

The Green New Deal, broadly speaking, involves large-scale public spending and investment in the promotion of renewable energy and electric cars. By creating stable, high-paying employment and therefore more effective demand, these measures are also meant to stimulate the economy. The expectation is that improved business conditions would lead to greater investment, which would spur the transition to a sustainable green economy.

This proposal reflects the wish for the return of the original New Deal that saved American capitalism after the Great Recession of the twentieth century. Neoliberalism has already proven to be ineffective in the context of the Anthropocene. Austerity and "small government" are unable to address the chronic outbreak of emergencies like pandemics, wars, and climate change. Now is the time for a new environmental Keynesianism—a "green Keynesianism"— to take their place. Its active and expansionary fiscal policy aims to accelerate investments in green infrastructure.

But can such a lovely story be true? Can a Green New Deal really save us from the Anthropocene?

THE BUSINESS OPPORTUNITY OF "GREEN ECONOMIC GROWTH"

Among those championing the Green New Deal, the person with the most heightened expectations for the economic growth it promises is the *New York Times* journalist Thomas Friedman. It also has to be framed as an opportunity and in our case as the most important opportunity for American renewal.[28]

Friedman had long argued that the wave of globalization following the collapse of the Soviet Union and advances in information technologies have led to a "flattening" of the Earth in which everyone is more connected than ever. In his 2008 book, *Hot, Flat, and Crowded*, he argues that by adding a "green revolution" to the mix, this flat world can achieve true sustainability.

As one can see from the quote above, green Keynesianism offers the hope that climate change might give us a golden opportunity for fostering more economic growth than ever before. Phrased another way, the "green economic growth" promised by green Keynesianism is the "last stand" for capitalism's ability to "function normally."

SDGS—IS UNLIMITED GROWTH POSSIBLE AFTER ALL?

The most prominent symbols of capitalism's last stand are the Sustainable Development Goals, or SDGs. The United Nations, the World Bank, the IMF (International Monetary Fund), the OECD (Organization for Economic Cooperation and Development), and other international organizations have all promoted SDGs as part of their passionate calls for green economic development.

For example, a seven-nation initiative (including South Korea and the United Kingdom) called the Global Commission on the Economy and Climate has created *The New Climate Economy Report*. In its 2018 edition, this report heaps praise on SDGs, claiming that sustainable development will be "driven by the interaction between rapid technological change, sustainable infrastructure investment, and increased resource productivity." As its authors put it, "We are on the cusp of a new economic era."[29] It's clear that the elites who populate international organizations see climate change solutions as golden opportunities for renewed economic growth.

The truth is, the green Keynesianism championed by pundits like Friedman and Rifkin would indeed spur further economic growth. The widespread use of solar panels, electric cars, and quick-charging batteries, along with the development of biomass energy, will necessitate large-scale shifts in the economy, leading to substantially increased financial investment and job creation. Moreover, it's undeniably true that existing social infrastructure will have to be completely overhauled in the age of climate crises, which will involve large-scale investment as well.

However, problems still remain. One cannot help but wonder: Is this type of growth really compatible with the planet's limitations? Does sticking a "green" label onto the hunger for unlimited growth really guarantee that it won't eventually overrun the ability of the planet to satisfy it?

PLANETARY BOUNDARIES

The fact is, even if we decide to foster economic growth to counter climate change, there are limits we must not exceed in terms of the amount of additional environmental burden produced by the great transition to a sustainable economy. This is the opinion of the environmentalist scholar Johan Rockström. He and his research team presented the concept of "planetary boundaries" in 2009.

First let me explain the basic idea.

The planetary system is supported by the innate resilience of nature. But if burdens above a certain limit are placed on this system, this resilience is lost, and the possibility arises that abrupt, irreversible, and destructive changes will occur, such as the melting of the polar ice caps or the mass extinction of plants and animals. These are known as tipping points. Perhaps it goes without saying, but passing these tipping points would spell disaster for humanity as well.

Rockström has defined the limits within which humanity can continue to enjoy a stable existence, mapping out the thresholds demarcating nine sectors of planetary resilience. These nine sectors include climate change, loss of biodiversity, biochemical flow of nitrogen and phosphorous, land-system changes, freshwater use, ocean acidification, ozone layer depletion, atmospheric aerosol loading, and pollution by chemical substances.

These are the planetary boundaries. Rockström proposed a plan by which "humanity can continue to develop and thrive" in a stable manner as long as these boundaries are respected.

The concept of planetary boundaries has naturally had a strong influence on the formulation of SDGs. They became the target values driving technological innovation and optimization.

CAN CARBON DIOXIDE EMISSIONS BE REDUCED AS THE ECONOMY GROWS?

The problem is that according to Rockström's calculations, four planetary boundaries, including climate change and loss of biodiversity, have already been exceeded due to human economic activity.[30]

This is a well-demonstrated fact. As a result of humanity's attempts to conquer nature, the Earth's environment has undergone major irreversible changes already. This has placed humanity on a path toward a critically dangerous point after which nothing can be done. Is this a situation in which the "green economic growth" promoted by "green Keynesianism" should really be our goal?

I want to draw attention to one of Rockström's newspaper articles, which he published in 2019. The title of this article, published ten years after the presentation of the planetary boundaries paradigm, is provocative: "Green Growth Is Wishful Thinking: We Must Act."[31]

Up until then, Rockström had, like many of his fellow researchers, argued that the goal of keeping global temperatures from rising less than 34.7°F was reachable if a version of green economic growth that remained within the planetary boundaries he'd defined was put in place. However, in this article, that optimism is nowhere to be seen. Rather, he proposes to the public that it must choose between two mutually exclusive actions: continuing economic growth or keeping global temperatures from rising more than 34.7°F. To put it in slightly more technical language, Rockström concludes that decoupling economic growth, even when adjusted to meet the 34.7°F target, and the environmental burdens of such growth is, in reality, extremely difficult.

WHAT IS DECOUPLING?

"Decoupling" might be a word seldom encountered in daily life, but it's a term used widely in the fields of both economics and environmentalism.

Normally, economic growth increases burdens on the environment. Decoupling is the attempt, through new technologies, to sever the link between economic growth and increases in environmental burden. In other words, it's an effort to find ways to grow the economy without worsening the impact of that growth on the environment. In the case of climate change, this means developing technologies that would support economic growth while reducing carbon dioxide emissions.

For example, the large-scale consumption of cars and housing and the creation of infrastructure like power plants and grids as Global North countries develop spurs their economic growth, but it also leads to huge increases in their carbon dioxide emissions. But if these countries are able to support this development by adopting more efficient new technologies, the volume of carbon dioxide emissions will rise in a gentler curve than if this infrastructural shift and large-scale consumption occurred while relying on old technology. The classic form this takes is the introduction of energy-saving technologies, hybrid cars, natural gas power generation, and the like.

Lessening the increase of carbon dioxide emissions relative to what would normally accompany increased economic growth by optimizing technological efficiency is referred to as "relative decoupling."

THE NEED FOR ABSOLUTE REDUCTION OF EMISSIONS

Relative decoupling is, unfortunately, woefully inadequate as a measure to combat climate change. The rise in global temperatures cannot be curbed unless carbon dioxide emissions are reduced *absolutely*, not relatively. The attempt to grow economically while reducing *absolute* emissions of carbon dioxide is known as "absolute decoupling."

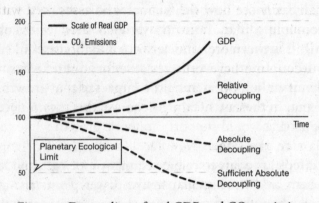

Figure 3. Decoupling of real GDP and CO2 emissions
Based on Kate Raworth, *Doughnut Economics: Seven Ways to Think Like a
21st-Century Economist* (New York: Random House, 2017)

Figure 3 fixes the real GDP and the amount of carbon dioxide emissions at one point in time at one hundred, then displays the subsequent shifts relative to that point. Looking at it, we can see how overwhelmingly different the amount of emissions that need to be reduced are in the cases of absolute decoupling and relative decoupling.

One example of absolute decoupling is the widespread adoption of electric cars, which emit no carbon dioxide at all. As the number of petrol-burning cars decreases, so does the amount of carbon dioxide emissions, while the sale of electric cars allows economic growth to continue.

Another example of absolute decoupling is conducting business meetings online via telecommunications technology rather than requiring people to board airplanes and go on business trips. The shift from fossil fuel power generation to solar power is also an example; growth continues even as emissions decrease. The idea is to minimize the relationship binding economic growth to increased emissions with the aim of severing it completely. If we combined multiple measures of this sort, it would be possible for the economy to grow while carbon dioxide emissions decreased in *absolute*, not relative, terms.

One can easily see how the technologies associated with relative decoupling differ from those associated with absolute decoupling. Furthermore, the technologies that would enable absolute decoupling have yet to be widely adopted. This is precisely why their adoption requires large-scale investment, and also why they represent such a golden opportunity for fostering economic growth.

In this way, the Green New Deal trumpeted by people like Thomas Friedman aims to reduce absolute carbon dioxide emissions to zero and keep global temperatures from rising more than 34.7°F while allowing the GDP to grow in the same manner it always has. Of course, this would necessitate a correspondingly large-scale technological revolution. The Green New Deal is this century's grandest project to bring this about and thus effect an absolute decoupling of economic growth and environmental damage.

THE GROWTH TRAP

When contemplating the possibility of future technological revolutions, we can imagine renewable energy and information technologies advancing at a pretty high speed. This is why several optimistic environmental economists proclaim that "absolute decoupling will be relatively simple."[32]

But will bringing about absolute decoupling really be so easy?

To answer this, we must ask, when will we be able to fully convert to a carbon-free society? It seems possible to do this, and thus reach the zero-emissions goal, within the next one hundred years.

But by then it will be too late. We need to remember the warnings of scientists—that we need to reduce emissions by half by 2030 and entirely by 2050. In other words, humanity's destiny lies in whether we will be able to effect *sufficient* absolute decoupling to stop climate change within the next ten to twenty years.

At this point, even someone like Rockström has accepted that the idea of fostering green economic growth through decoupling is nothing more than "wishful thinking." It's impossible to effect an absolute decoupling of sufficient scale to reach the goal of keeping global temperature increases below the 34.7°F mark.

But why is it impossible? The answer lies in the simple yet intractable dilemma that haunts any decoupling attempt. The dilemma is this: as the economy grows, the range of human economic activity grows too, which means that the volume of resource and energy consumption also grows, making it difficult to reduce carbon dioxide emissions. This is a historical tendency.

In other words, even green economic growth may cause increases in carbon emissions and resource use in direct proportion to its success because economic growth is historically accompanied by more frequent consumption of bigger commodities, including ones in wasteful and carbon-intensive industries. This in turn will necessitate more and more dramatic increases in efficiency, but there is an insurmountable physical limit to the improvement of technological efficiency. This is the Growth Trap, a major pitfall awaiting capitalism as it attempts to establish a zero-carbon economy. The question is, can this trap be avoided?

To skip to the conclusion, it seems that, unfortunately, escaping this trap is unlikely. Sustaining a GDP growth rate of 2–3 percent would necessitate the immediate reduction of carbon dioxide emissions by 10 percent every year in order to hit the 34.7°F target. If we leave it to the market, the likelihood of achieving a yearly reduction rate as dramatic as 10 percent is very low.

THE PRODUCTIVITY TRAP

Rockström states that once we face the Growth Trap directly, the natural solution that comes to the fore is giving up on economic growth. The reason is simple: if we give up on growth and allow the scope of the economy to contract, reaching our carbon dioxide emission reduction goals will become easier.

This is the one decision we must make if we want to halt the destruction of the global environment and sustain the conditions necessary for humanity to thrive. But it's a decision that cannot be made under capitalism. Here we encounter another of capitalism's traps, the Productivity Trap.[33]

Capitalism is always trying to raise workforce productivity in order to cut costs. Rises in workforce productivity allow the same amount of production to occur with fewer workers. When this happens, the economy's size remains constant while unemployment rates rise. But capitalism also makes it impossible for the unemployed to live, and politicians hate high unemployment rates. For this reason, there's a huge amount of pressure for the economy to keep expanding indefinitely so as to maintain the rate of employment. This is why a rise in productivity results in the expansion of the economy. This is the Productivity Trap.

Capitalism cannot escape the Productivity Trap, which means it cannot wean itself off its dependence on economic growth. This is why even if we were to put in place measures to combat climate change, we would end up falling into the Growth Trap, resulting in further increases in resource consumption.

This is also why scientists are beginning to sense capitalism's limits.

DECOUPLING IS AN ILLUSION

The conclusion that we should give up growth likely seems rather unreasonable to readers who've followed my argument so far. Green Keynesianism certainly sounds more commonsensical, and economic growth seems like something that must never be abandoned. This is why I would like to introduce some empirical research examining why decoupling is so difficult to do. One study I'd like to look at was conducted by the famous British environmental economist Tim Jackson and presented in his bestselling book, *Prosperity without Growth* (2009, second edition 2017).

Jackson points out that improving the efficiency of energy consumption has already been a central focus of the developed world's industrial sector. In the US and the UK, there has been an impressive

40 percent improvement in this area since the 1980s. Looking beyond those examples and focusing on countries belonging to the OECD, the percentage of their real GDP taken up by energy consumption has significantly decreased. In short, it appears that relative decoupling is proceeding apace—that is, as long as you look only at the Global North.

However, in Global South countries like Brazil and those in the Middle East, the trend is the reverse—in those countries, the percentage of the GDP taken up by energy consumption is rapidly increasing. As short-term economic growth takes priority, large-scale investments are occurring without updated technology, creating an environment in which not even relative decoupling can occur.

If energy consumption efficiency doesn't improve, then naturally the proportion of the real GDP taken up by carbon dioxide emissions cannot improve either. Because the center of world economic growth has shifted to China and Brazil, worldwide emissions have decreased only by a scant 0.2 percent between 2004 and 2015.[34]

In other words, when viewed on a global scale, recent years have shown us that even the relative decoupling of carbon dioxide emissions and economic growth has barely occurred. The sufficient *absolute* decoupling needed to reach the 2050 zero-emissions target remains nothing more than a dream.

It's true, though, that in many Global North countries, the long-term economic stagnation that followed the Lehman Brothers bankruptcy in 2008 contributed to a reduction in carbon dioxide emissions. For example, in the UK, the GDP rose by 27 percent between 2000 and 2013 while emissions decreased by 9 percent. Absolute decoupling was also observed in Germany and Denmark during this time.

Again, though, looking at things on a global scale, emissions are still going up due to the rapid economic growth occurring in emerging nations. So the reality is that even as some carbon dioxide emissions are decreasing as a result of absolute decoupling, overall emissions are still on the rise. This data conforms to what we've already seen (see Figure 2). Overall, the worldwide volume of carbon dioxide emissions is rising by about 2.6 percent every year. Even emissions in the US are rising at a yearly rate of 1.6 percent.[35] It's simply not realistic to think that the sufficient absolute decoupling needed to reach the under 28.4°F mark will happen if things continue this way.

This is why Jackson concludes that decoupling is a "myth" and criticizes the arguments of those promoting green economic growth as "totally unconvincing." Further, he calls the idea that technological advances under capitalism will combat climate change a "simplistic assumption" that's really no more than an "illusion."[36]

WHAT'S REALLY HAPPENING IS RECOUPLING

Looking at the data Jackson presents, some might react by saying that emerging nations should be punished, since the continued rise in worldwide carbon dioxide emissions is largely due to rapid economic growth in the Global South.

However, this would amount to nothing more than a repeat of the Netherlands Fallacy discussed in the previous chapter. It's misleading to focus only on the reduction of emissions occurring in some Global North countries. After all, the resources being extracted and the goods being produced in developing markets like China, Brazil, and India are in no small part bound for export and consumption in the Global North.

In other words, the apparent decoupling occurring in some developed countries is in fact simply the result of environmental burdens (in this case, the carbon dioxide emissions produced by economic activity) being displaced onto an exterior. The decoupling achieved by OECD nations has resulted not only from advances in technology but also from the displacement of the production of consumer goods and foodstuffs consumed by those nations to the Global South over the past thirty years.

On top of this, Jackson argues that if we factor in the additional carbon footprint involved in importing these products, the apparent relative decoupling disappears entirely.[37] The "carbon footprint" here refers to the amount of greenhouse gasses produced during the entire process of creating and consuming a product or service, from the procurement of raw materials to the product's eventual disposal as waste, added to the calculation of total carbon emissions..

Because of this, while absolute decoupling seems possible in theory, in practice, the likelihood that it could occur on a large scale over a sustained period of time—that is, outside of temporary emergencies and financial downturns like the COVID-19 pandemic or the 2008 financial crisis—is extremely low.

Fundamentally, no matter how much technology may advance, efficiency and optimization have physical and thermodynamic limits. No matter how efficient things get, we will never be able to create cars using half the amount of resources we do now, and creating storage batteries and electric cars takes energy as well.

As a quick look at the history of capitalism since the Industrial Revolution tells us, economic growth during the twentieth century was only possible through the use of enormous amounts of fossil fuels. Economic growth and fossil fuel consumption are intimately and inextricably linked, and thus fossil fuel cannot be replaced with green energy. It's physically impossible to sustain the same level of economic growth as before and reduce carbon dioxide emissions at the same time.

Taking this into account, it would be a mistake, in a time of climate crisis, to place our hopes on a form of economic growth dependent on absolute decoupling. This is precisely why "green growth" strategies for combating climate change, which spread the illusory notion that absolute decoupling is a simple thing to achieve, are so dangerous.

INCREASING EFFICIENCY INCREASES ENVIRONMENTAL BURDEN—THE JEVONS PARADOX

There is another inconvenient truth we must consider. A key component of decoupling is increasing efficiency, but paradoxically, increasing efficiency makes combating the climate crisis more difficult.

For example, there is at present increased investment in renewable energy all around the world. Yet fossil fuel consumption is not decreasing. This is because renewable energy isn't being consumed in place of fossil fuels but rather *alongside* fossil fuels as overall energy demands rise due to economic growth.

Why is this happening? One explanation may be found by applying the Jevons Paradox. This is the name given to an effect described by the nineteenth-century economist William Stanley Jevons in his book, *The Coal Question* (1865).

In England at the time Jevons was writing, technological advances had greatly improved how efficiently coal could be used. But this didn't result in a decrease in the amount of coal being consumed. Rather, the drop in the price of coal due to how little was now needed resulted in it being used in all sorts of ways it hadn't been before, leading to an overall increase in its consumption. At this early point in history, Jevons was already pointing out how, contrary to the commonsensical assumption that improving efficiency will lead to a reduced burden on the environment, technological advancements of this sort in fact lead to an *increased* burden.

And indeed, the same thing is happening right now. Even as the development of new technologies is improving efficiency, the resulting decrease in the price of goods frequently leads to the increased consumption of those goods. A television might use less energy than before, but the increase in demand from people purchasing more and larger televisions leads to an increase in the overall amount of electricity consumed by televisions anyway. Another example is the rise in popularity of larger, gas-guzzling vehicles like SUVs—they render any gains in fuel efficiency meaningless due to the same principle. No matter how much apparent relative decoupling might occur due to the increased efficiency of new technologies, any such effect will end up being erased by the boomerang effect of increased consumption, rendering the whole effort meaningless.

Furthermore, even if relative decoupling occurs in one area due to increased efficiency, the capital and income gained from that efficiency frequently ends up applied to the production or purchase of other goods that use up more energy and resources, rendering any gains from efficiency moot. An example of this is people buying cheaper household solar panels, only to use the money they save to buy larger gas-powered vehicles. Companies will always find a way to reinvest any surplus capital they produce, and there's no guarantee this investment will be green.

In these cases, relative decoupling in one area leads, ironically enough, to the obstruction of absolute decoupling overall.

THE POWER OF THE MARKET CANNOT STOP CLIMATE CHANGE

Here I would like to point out a different problem with the green Keynesianism promoted by Rifkin and others. Green Keynesianism aims only to stimulate the market, not regulate it in any way. But the market's price mechanism can never function as a way to reduce carbon dioxide emissions.

To illustrate this failing of the market, let us look at the phenomenon of "peak oil." The concern is that if oil production passes its peak, supply will go down and the price of crude oil will go up, which might have a negative effect on the economy. There have been repeated debates about when peak oil will be reached, as well as exactly what effects it will have on the world economy.

Market fundamentalists look at the problem in the following way. If there's a sharp rise in oil prices, new technologies like renewable energy will become correspondingly cheaper. The cheaper it gets, the more renewable energy development will advance. As a result, oil consumption will naturally go down.

In reality, though, things work differently. Under capitalism, a rise in oil prices simply means that previously unprofitable oil sources, such as oil sands and shales that need to be upgraded into synthetic crude, are exploited. Companies convert sharp rises in prices into more opportunities to make money.

Some might object that even if this is true, there will come a time when future innovations will progress to the point that renewable energy is so cheap that oil will no longer be profitable at all. Indeed, none other than Jeremy Rifkin argues passionately that the market's price mechanism will make the "collapse of fossil fuel civilization" all but inevitable.[38]

However, when confronted with a hypothetically rapid development in renewable energy technology that would make oil lose its price-competitiveness, is it really the case that the oil industry would respond by putting itself out of business? Of course not. Instead, the oil industry would attempt to extract as much fossil fuel from the Earth as it could while there was still a possibility to sell it, only leading to a rise in the pace of extraction. These would be its final death throes.

This is a terribly dangerous, even fatal error to make in the face of an irreversible problem like climate change. It's also why the compelling force necessary to effectively reduce greenhouse gas emissions must come from outside the free market.

THE HUGE AMOUNT OF CARBON DIOXIDE EMITTED BY THE RICH

Whatever the case, the fact that wide-ranging, consistently applied decoupling is extremely difficult to bring about means that the promises made by green Keynesianism can never be kept. Even if politicians are able to win elections on the grandiose platforms of Green New Deals, they will never be able to fulfill their promises to solve the environmental crisis.

The problem has deeper roots than that. In short, the large-scale production and consumption that fueled economic growth up until now must itself be radically rethought. This is why more than 1,100 scientists issued a statement in 2019 pointing out that the "climate crisis is closely linked to excessive consumption of the wealthy lifestyle" and calling for the radical transformation of the mechanisms by which the present economy functions.[39]

Of course, the wealthy lifestyle of which they speak is that of the rich populations of the Global North, who are responsible for an excessive proportion of carbon dioxide emissions. A shocking piece of data shows that the carbon dioxide emitted by the top 10 percent of the world's richest people makes up half of worldwide emissions. The top 0.1 percent especially has an extremely serious impact on the environment as they drive around in sports cars, hop from place to place in private jets, and maintain multiple mansions.

On the other hand, the least wealthy half of the world's population is responsible for a mere 10 percent of worldwide carbon dioxide emissions. Yet these people are also the first to suffer from the effects of climate change. This is where we see the contradictions of the Imperial Mode of Living and the externalization society play out very clearly. And this is why it is perfectly appropriate to call for primarily the rich to reduce their emissions. It is a problem born from the Imperial Mode of Living.

The fact is, if the world's richest 10 percent were to lower the amount of emissions they produce to that of the average European, overall emissions would decrease by a full third.[40] This would likely buy sufficient time for a comprehensive transition to a sustainable social infrastructure.

But we must also keep in mind the following: almost every one of us living in a developed country belongs to the world's richest 20 percent, and some of those who call themselves "middle class" are actually part of the top 10 percent. In other words, it will be impossible to truly combat climate change if we all fail to participate as directly interested parties in the radical transformation of the Imperial Mode of Living.

THE TRUE COST OF THE ELECTRIC CAR

That said, what *would* happen if, despite all this, green investment increased with the aim of spurring economic growth, expanding markets with the idea that decoupling will work? Let's think about this using the example of an electric car like the Tesla.

There's no question that at present, gas-powered vehicles produce an enormous amount of carbon dioxide emissions worldwide. For this reason, it is imperative to introduce low-carbon modes of transportation, and nations must actively support their development.

The same goes for the conversion to energy-saving and renewable energy–consuming technologies.

As I've said previously, replacing all gas-powered cars with electric cars would open up enormous new markets and employment opportunities. This would, in theory, solve both the environmental crisis and the economic crisis. This is the dream of green Keynesianism taking its most ideal form.

But the real story isn't so sweet.

The key to this entire project is the lithium-ion battery that won Akira Yoshino the Nobel Prize in Chemistry in 2019. Lithium-ion batteries are indispensable for the functioning not only of smartphones and laptop computers but also of electric cars. The drawback is that large amounts of rare metals are needed to manufacture them.

The first of these is, naturally, lithium. Lithium deposits can be found in many regions along the Andes mountain range. Chile, for example, with its deposits in the Atacama Salt Flats, is a major lithium producer.

Lithium becomes concentrated in the groundwater beneath dry regions over a long period of time. Lithium-rich brine is pumped from beneath salt lakes, and then the lithium is extracted by evaporating the remaining water. Put simply, lithium extraction and groundwater drainage are one and the same.

The problem is one of volume. A single corporation can extract 1,700 liters of groundwater per second. The drainage of such a huge volume of water in an area that's already so dry will obviously have a great impact on the region's ecology.

For example, the population of Andean flamingos who depend on the shrimp who live in this briny water is decreasing. Further, the rapid drainage of groundwater is causing shortages in the fresh water accessible to residents of the area.[41] Lithium extraction in Argentina is reportedly causing a similar situation. Basically, the effort to combat climate change in the developed world is causing even more intense extraction and exploitation in the Global South to meet the demand for a different resource meant to replace oil. Thanks to spatial shift, this phenomenon is rendered invisible.

Another necessary element for manufacturing lithium-ion batteries is cobalt. The problem here is that almost 60 percent of the world's cobalt is mined in the Democratic Republic of Congo, one of Africa's poorest and least politically and socially stable nations.

Cobalt extraction is straightforward: cobalt deposits beneath the Earth's surface are mined using heavy equipment and human labor. It goes without saying that the large-scale mining necessary to meet worldwide demand, which is only continuing to expand, has led to environmental damage such as water pollution and the pollution of crops as well as the destruction of the landscape. On top of this, there is the problem of terrible labor conditions.

In the south of Congo, informal systems of child and slave labor are flourishing under the rubric of the *creuseur*—a French term frequently translated as "artisanal digger." Using primitive tools like hammers and chisels, *creuseurs* frequently mine for cobalt with their bare hands. Some of these workers are children as young as six or seven years old—there are forty thousand children working in this industry—each making around the equivalent of a single US dollar per day.

Furthermore, the mining is being conducted in dangerous tunnels that lack necessary safety features. It's not uncommon for people to spend more than twenty-four hours at a time underground, and prolonged exposure to toxic materials in the air leads to health problems such as heart and lung disease as well as psychological maladies.[42] In the worst cases, mining accidents lead to miners being buried alive. The deaths of children among these workers has led to international condemnation.

On the other side of this global supply chain is not only Tesla but Microsoft and Apple as well. It's not as if the heads of these major corporations are unaware of how the lithium and cobalt they need are procured. In fact, there's even a human rights organization in the US that has instigated legal action regarding the issue.[43] Nevertheless, these companies continue to argue for developing new technologies in order to meet the all-important SDGs, seemingly without the least concern.

ECOLOGICAL IMPERIALISM IN
THE ANTHROPOCENE

In the end, the effort to bring about "green economic growth" in the developed world is nothing more than a further shifting of social and environmental costs onto the periphery. As I touched on in chapter 1, this is the same sort of ecological imperialism we saw regarding the extraction of guano from the islands off the shore of Peru in the nineteenth century. It's now repeating itself once more in South America and Africa, dressed in the new clothing of rare metals extraction.

The problem is not confined to lithium and cobalt. The demand for iron, copper, and aluminum has also risen as the GDP has continued to grow. The consumption of products that require these resources has risen precipitously as well. A team of environmental scholars headed by Australian professor Thomas Wiedmann has developed a calculation called the "material footprint" (MF) to revise our understanding of the effects of international trade. The MF is a figure indexing the consumption of natural resources.[44]

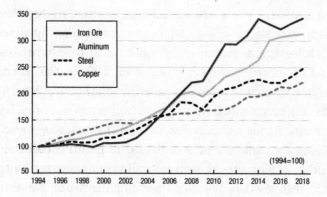

Figure 4. Increases in ore production
Based on data from US Geological Survey, National Minerals Information
Center, "Mineral Commodity Summaries," 1994–2019

According to the team's research, after this revision is performed, not even the MF of developed countries has been decoupled from their economic growth. While Domestic Material Consumption (DMC) might have fallen in some cases, once the MF of the resources used in importation is added in, it ends up taking up approximately the same proportion of the real GDP of every nation in the world. In other words, the relative and absolute decoupling occurring in the developed world is only ever temporary, and what has actually been happening is a *recoupling* of the GDP and its material footprint.[45]

The truth is, the total consumption of resources in 1970—including mineral resources, ores, fossil fuels, and biomass—was 26 billion 700 million metric tons, while in 2017 it surpassed 100 billion metric tons. By 2050, this figure is predicted to rise to 100 billion 800 million metric tons.

Only a mere 8.6 percent of these resources are recycled, a proportion that's actually decreasing in the face of the rapid increase in consumption. The "dematerialization of capitalism" driven by the developed world's recent shift to information and communications technology (ICT) and service industries may be trumpeted by some, but when we look at worldwide consumption of material resources, this dematerialization is nowhere to be found.[46]

However you slice it, it's clear that the present economic system is unsustainable. It's not simply that sufficient absolute decoupling is too difficult. The "circular economy" that some have put their hopes on to realize a sustainable society has proven to be a misleading idea as well. Recycling resources is, by itself, an insufficient solution. What must happen is a radical reduction in the actual volume of resource consumption.

The future for developed countries following the principles of green Keynesianism to chase capitalistic "green growth" is dark. It certainly may be the case that individual countries will adopt economic policies they proclaim to be "green." But the plunder of the periphery is only intensifying. So what sort of cart-before-the-horse version of green is this—green how, and for whom? The fact is that it's extremely difficult to avoid this kind of boomerang effect, as plunder itself is the basic ingredient necessary to protect the core's environment.

TECHNO-OPTIMISM IS NOT THE SOLUTION

On top of this, there is yet another inconvenient truth we must confront. The effectiveness of the green policies put into effect in some developed countries is doubtful. In places where households typically possess multiple cars and trucks, the result would still be unsustainable even if every single one were replaced by an electric vehicle. Furthermore, the planned rollout by Ford and Tesla of SUV-style electric vehicles signals nothing more than the continued strengthening of our present culture of consumption, which will only lead to an increased waste of resources. It is, when all is said and done, simply another classic example of greenwashing.

The production of electric cars involves the use of fossil fuels to extract the necessary raw materials, causing increases in carbon dioxide emissions and other types of environmental destruction. It's the Jevons Paradox all over again. In the end, the environmental crisis only gets worse.

There's another piece of data that puts the final nail in the coffin of the promise of the electric car. According to the International Energy Agency (IEA), by 2040, the number of electric cars is expected to rise from two million to more than 280 million, but this is expected to result in a mere 1 percent reduction in carbon dioxide emissions.[47]

How can this be? One reason is that as the number of electric cars grows, so too does the overall number of gas-powered vehicles worldwide, especially in the Global South. Furthermore, even if electric cars *were* to entirely replace gas-powered ones, carbon dioxide emissions wouldn't fall to zero. This is because the increased production of large-scale batteries would lead to increased emissions. The ever-increasing scale of both products and their production as fostered by capitalistic growth makes becoming carbon zero a near impossibility.

As the above has made clear, when we look at the production processes for green technologies, we can see that they are not very green at all.[48] The realities of production are rendered invisible, hiding how these solutions so frequently transform one kind of problem into another. Therefore, while the transition to electric cars and solar power is absolutely necessary, it would be a fatal mistake to stake our future on techno-optimism.

All that said, green Keynesianism's calls for a 100 percent transition to electric cars and renewable energy may still sound attractive. But this is only because green Keynesianism promises a sustainable future without having to change our Imperial Mode of Living—that is, without us having to do much of anything at all. To borrow Rockström's phrase, this is nothing more than wishful thinking.

CAN NEW TECHNOLOGIES REMOVE CARBON DIOXIDE FROM THE ATMOSPHERE?

If we cannot hope to reduce carbon dioxide emissions through the widespread use of electric cars, the only thing left for the "green growth" contingent to do is to place their hopes on ever-more-amazing technologies. If reducing emissions is too difficult, why not concentrate on developing technologies that would remove carbon dioxide from the atmosphere instead? These technologies are referred to as Negative Emissions Technologies (NETs), as they aim to reduce emissions to *less* than zero.

With NETs, absolute decoupling becomes much easier. The IPCC *Special Report on Global Warming of 1.5°C* put out by the UN in 2018 even incorporates the adoption of NETs into its projected scenario for keeping global temperature rises between 34.7° and 35.6°F. NETs are a shining star in the firmament of green Keynesianism.

As many climate scientists have pointed out, however, there are many problems with the IPCC's scenario and its assumptions about

NETs. The feasibility of creating and using NETs is hardly certain, for one thing, and even if they are put to use, the unwanted side effects are predicted to be serious.[49]

Let's look for a moment at the preeminent example of an NET, Bio-Energy with Carbon Capture and Storage (BECCS). BECCS proposes the adoption of biomass energy to reduce carbon dioxide emissions to zero along with technology that would remove carbon dioxide from the atmosphere and store it underground or in the ocean, bringing those levels down to less than zero.

But even if BECCS were put into practice, would the problem really be solved so easily? For one thing, the continued march of green economic growth would still demand the scope of economic activity to keep expanding, which means that the scale of BECCS would have to keep expanding along with it.

Looking at the biomass energy (BE) side of things, one problem that jumps out is the enormous amount of land needed to produce it. Farmland twice the size of India would be necessary to produce enough biomass energy to hit the target of keeping global temperatures from rising more than 35.6°F. How could that much land be secured? Would the burden be pushed upon places like India and Brazil once again, forcing locals to give up land they need to cultivate food? Or would the land be cleared by slashing the Amazon rainforests even further, thus rendering any reduction in carbon dioxide emissions moot?

The CCS—carbon capture and storage—side of things presents problems as well. The biomass power plants associated with this process use enormous amounts of water; it's projected that it would take 400 million metric tons of water to produce enough electricity to power the US for a year. We are already having problems with the amount of water being used to produce crops now, and as climate change progresses, water will only become more precious. How can we contemplate investing in a new technology that would use it in such great quantities? Furthermore, if CSS deposits large amounts of carbon dioxide into the ocean floor, it will become even harder to combat the already widespread problem of ocean acidification.

But the biggest issue is that CSS probably won't even work. There's a significant possibility that large volumes of carbon dioxide deposited beneath the Earth's surface will end up leaking back out. Odorless and colorless, the presence of this leaking carbon dioxide could easily be denied by industry until it is too late.

In short, BECCS is a technological fix that consists of nothing but all the displacements identified by Marx writ large.

THE "INTELLECTUAL GAME" PLAYED BY THE IPCC

There is something worth noting at this juncture. What would be the purpose of using other natural resources in such huge amounts, thus increasing the overall burden on the environment, in order to enable us to keep using fossil fuels anyway? Wouldn't we do better to create a society that's no longer dependent on fossil fuels at all? There's no angle from which BECCS does not prove itself to be a terrible solution to climate change.

Yet the Fifth Assessment Report (AR5) put out by the IPCC includes "dream" technologies like BECCS in almost all their scenarios for keeping global warming below 35.6°F. The experts drawing up these reports must be well aware of how flawed BECCS is, but they keep mentioning these unrealistic processes as they construct the steady stream of complicated models and scenarios they present to the world.

There's no way to avoid seeing it as a simple "intellectual game" for scholars, to quote Rockström's criticism. Shouldn't these top-level experts be explaining to the public that something must be done to head off the impending crisis and telling politicians and government officials why more comprehensive measures must be taken immediately rather than spending their precious time cooking up these kinds of unrealistic dream scenarios?

There are likely those who, upon hearing these criticisms, will be mystified as to why the IPCC has descended into such self-contradiction and navel-gazing. The reason, though, is simple. The IPCC model is premised on economic growth and is thus unable to escape the Growth Trap. As long as this is true, the only solutions they can see are dependent on technologies like NETs.

THE ROAD TO EXTINCTION IS PAVED WITH GOOD INTENTIONS

As demonstrated above, the introduction of electric cars and the transition to renewable energy are things that must be done, but if we do them only as ways to allow us to keep our current lifestyles the way they are, we will easily find ourselves captured by the logic of capital and ensnared in the Growth Trap.

The only way to avoid this trap is to disengage from a consumer culture that equates car ownership with independence while also reducing the volume of *everything* we consume. We must make a major incision into capitalism itself to heal the world. This is why green Keynesianism is not enough.

Make no mistake—Green New Deal–style governmental platforms enabling large-scale investment into remaking nations at a fundamental level are absolutely indispensable in the struggle to combat climate change. It's undeniable that we must make the transition to solar energy, electric vehicles, and the like. Public transportation systems must be expanded and made free to all, bicycle lanes must be built, and public housing fitted with solar panels must be created—these sorts of works projects, driven by public spending, are all vital.

But these things are simply not enough. It might sound counterintuitive, but the goal of any Green New Deal should not be economic growth but rather the scaling down and slowing down of the economy.

Measures to stop climate change cannot double as ways to further economic growth. These measures will only work if their only goal is stopping climate change. Indeed, the less such measures aim to grow the economy, the higher the possibility is that they'll work. Slowing down the economy also would ease the issues related to lithium and cobalt extraction in Chile and Congo (though environmental problems will nonetheless still surely occur).

By contrast, the only thing left to say at this point about Green New Deal proposals aiming to promote unlimited economic growth is, "The road to extinction is paved with good intentions."[50]

THE MYTH OF DEMATERIALIZATION

Such a proclamation may grate on many readers. But I am not alone—there are many scholars and researchers who have come to similar conclusions in recent years. For example, Vaclav Smil, a Canadian professor reportedly beloved by Bill Gates, clarified his position thus in his 2019 book, *Growth*: "Continuous material growth [. . .] is impossible. Dematerialization, doing more with less—cannot remove this constraint."[51]

As Smil points out, the so-called "dematerialization" brought about by the shift to a service economy will not solve the problem. For example, while leisure is not itself material, the carbon footprint of leisure activities makes up 25 percent of the overall total.[52]

The vaunted "internet of things" (IoT)—the term coined by Jeremy Rifkin—used to further develop the information economy is not a solution either. Contemporary capitalism may appear to be creating an economic system prioritizing mental labor and the dematerialization of society, but in truth, the manufacture and operation of computers and servers consumes enormous amounts of energy and resources. The so-called "cognitive capitalism" dependent on information and communications technologies is a far cry from something that could bring about true dematerialization or decoupling. It's just another myth.

Pundits like Rifkin and Friedman offer no persuasive answers to such critiques of their proposals. They simply remain silent when confronted with inconvenient truths, preferring to trumpet the merits of their solutions instead.

IS STOPPING CLIMATE CHANGE IMPOSSIBLE?

Looking at it this way, we may well start to wonder if those calling for a Green New Deal truly believe in stopping climate change at all. There are, for example, Green New Deals that focus not on "halting" or "relieving" climate change but rather "adapting" to a world that has warmed more than 37.4°F as they foster economic growth. These adaptation strategies come with proposals promoting NETs, nuclear energy, geoengineering, and the like.

These are the exact sorts of proposals put forward, for example, by the famous American environmental think tank the Breakthrough Institute, and they form a point of commonality between other prominent figures privileging "adaptation" to climate change, such as Steven Pinker and Bill Gates.

Suggesting we "adapt" to climate change is nothing more than a way to say that climate change cannot be stopped. Isn't it a bit early to give up while the possibility that we can combat it remains? Shouldn't we first do everything we can, while we still can?

One of the reference points for imagining what this sort of "adaptation" would entail is a return to the level of lifestyle available in the latter half of the 1970s.[53] People living in Asia and Europe, for example, would no longer be able to take planes just to spend a few days in New York City. People would no longer be able to drink Beaujolais Nouveau flown in on the same day as its release. But what effect would such minor restrictions have? Compared to what will happen if global warming reaches 37.4°F, these adjustments amount to nearly nothing. In the latter case, there would be no French wine to drink at all, as it would no longer be possible to cultivate grapes in France.

Of course, we all know that these harsh visions of how much our lives might deteriorate in the future hardly make for attractive political choices. But to respond to this difficulty by turning our backs on it and supporting more palatable "green growth" political packages in order to win elections, no matter how sincere the intentions behind them might be, amounts to nothing more than greenwashing dressed up as concern for the environment.

It's exactly this sort of wishful thinking that further strengthens our attachment to the Imperial Mode of Living, resulting in more pressure to exploit the periphery than ever. If we continue down this path, it won't be long before we end up suffering the consequences at home as well.

CHOOSING DEGROWTH

Giving up on the wishful thinking of green economic growth entails making a series of hard choices. How serious are we about reducing carbon dioxide emissions? Who will shoulder the cost? What sort of reparations are we willing to make to the Global South for everything taken from it by the Imperial Mode of Living? What are we prepared to do about the additional environmental destruction caused by the very process of transitioning to a sustainable economy?

There are no easy answers to these questions. However, one possible answer is *degrowth*. Obviously, choosing the path of degrowth would not solve everything, and we may still fail to change things enough or to do it soon enough. Starting with the next chapter, however, I want to demonstrate that degrowth is an idea that we cannot write off if we are serious about preventing the worst-case scenario from coming to pass.

Naturally, one major issue to consider at this point is exactly what sort of degrowth we should aim for. This will be the subject of the next chapter.

3

Shooting Down
Degrowth Capitalism

FROM ECONOMIC GROWTH
TO DEGROWTH

We just saw how reducing carbon dioxide emissions with sufficient speed to head off the climate crisis is impossible if we pursue economic growth at the same time. Decoupling is simply too difficult. Therefore, we have no choice but to abandon economic growth and consider degrowth as the foundation for any plan to combat climate change. But what sort of shape should this degrowth take? This chapter will investigate possible answers to that question.

But before we get to that, there's something I feel I must clarify. There are millions of people all over the world who have no access to electricity or potable water, to education or even sufficient food. For these people, economic development is obviously necessary.

In the field of developmental economics, people have consistently pointed to economic development as the key to solving the problem of the North-South Divide and have attempted to do so in various ways. I would never deny the benevolence and importance of these efforts.

But it's also true that development models with economic growth as their focus are reaching an impasse. Criticism of the World Bank and the International Monetary Fund (IMF) has also been growing.[54]

One such critic to whom I'd like to turn our attention is the political economist Kate Raworth. After tackling the problem of the North-South Divide for many years at Oxfam, an NGO focusing on aiding development internationally, she began criticizing mainstream economics and supporting degrowth.

As a step toward figuring out what sort of degrowth will be necessary to survive the Anthropocene, let's first to listen to what Raworth has to say.

DOUGHNUT ECONOMICS—THE SOCIAL FOUNDATION AND THE ECOLOGICAL CEILING

The starting point for Raworth is the question, "How much economic development can occur that will allow all of humanity to thrive while respecting the planet's ecological limits?" To answer this question, Raworth developed the concept of "doughnut economics."

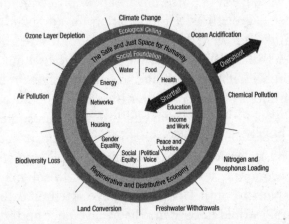

Figure 5. Conceptual diagram of doughnut economics
Based on Kate Raworth, *Doughnut Economics: Seven Ways to Think Like a 21st-Century Economist* (New York: Random House, 2017)

As a quick glance at the graphic shows us, the inner circle of the "doughnut" is formed by a "social foundation" while the outer circle represents the "ecological ceiling."

Her first point is that as long as there are people whose lives lack a basic social foundation—access to adequate water, shelter, education, etc.—then we cannot say that humanity is thriving. Those who lack a social foundation have insufficient material conditions to realize their full potential to live well. If the latent potential of humanity isn't allowed to bloom, a just society can never come into being. This is the situation in which so many of those living in the developing world find themselves.

Realizing humanity's full potential is not a matter of simply allowing everyone to behave in any manner they wish. Sustainability is necessary to ensure that future generations will also have the chance to thrive. This is where the outer ring of the doughnut comes into play—the ecological ceiling, based on the planetary boundaries defined by Johan Rockström that we discussed in chapter 2.

In short, Raworth's fundamental assertion is that a sustainable and just society can be realized only if a global economic system is put in place that will allow as many people as possible to thrive in the space between these upper and lower boundaries.[55]

However, as we've seen again and again thus far, those of us living in developed countries do so in a way that already grossly overruns multiple planetary boundaries while at the same time, many of those in the Global South live at a level below Raworth's definition of the minimum social foundation. The present system is not only hugely destructive to the environment but profoundly unjust as well.

WHAT IS NEEDED TO RECTIFY INJUSTICE?

Raworth's framing of the issue has had an enormous impact, inspiring not only research in the area of political science but across other disciplines as well. One example of this is the empirical study done by British environmental economist Daniel O'Neill.[56] O'Neill's research uses Raworth's doughnut economics model to concretely define the living standards and environmental burdens of approximately 150 countries around the world, measuring how well each fits within the inner and outer rings of Raworth's doughnut.

When O'Neill examined quality of life relative to environmental damage, his research proved that the more stable a nation's social foundation was, the greater the tendency for that nation to overshoot planetary boundaries. Almost every nation satisfied social demands by sacrificing sustainability.

This is an incredibly inconvenient truth to uncover. It means that using developed nations as models when helping emerging countries raise their living standards to attain the minimum social foundation will inevitably, when seen from a planetary point of view, lead us down the path to total destruction.

According to Raworth, however, even if some increased resource and energy consumption becomes necessary, the additional burden produced by rectifying social injustice is much lower than is commonly assumed. Using food as an example, increasing the world's overall caloric intake by just 1 percent would save 800,060,000 people from starvation. As another example, around 1.3 billion people do not have access to electricity, but providing them with it would increase overall carbon dioxide emissions by only around 1 percent. It would take the redistribution of a mere 0.2 percent of the world's wealth to end the hardship of the 1.4 billion people currently living beneath the world poverty line of US $1.25 a day.[57]

Moreover, though Raworth doesn't address this directly, promoting democracy and gender equality produces no environmental burden at all. Economic equality, too, if realized via the redistribution of subsidies currently spent on the fossil fuel industry ($5.9 trillion, or 6.8 percent of the GDP in 2020), would produce no additional environmental burden. In fact, it would likely improve the environment!

As this line of research shows, the grossly unjust divide between the Global North and South could be largely rectified without further damaging the environment, provided we stop clinging to the dream of perpetual economic growth.

ARE ECONOMIC GROWTH AND HAPPINESS REALLY CORRELATED?

Another important point made by Raworth is that past a certain level, the link between economic growth and improving people's lives becomes difficult to discern. Once basic standards are met, it's not at all clear that the premise is true that only economic growth can cause society to thrive.

This is easy to see when comparing Europe to the US. The per capita GDP of most northern European nations like France and Germany is lower than that of the United States. But their standards of social welfare are much higher, and many of these nations provide healthcare and higher education free to their citizens. In the US, by contrast, some people lack health insurance and therefore have difficulties accessing healthcare, and many people struggle with student loans they will never be able to pay back. Japan's per capita GDP is also much lower than America's, but the average Japanese lifespan is almost six years longer.[58]

In other words, the extent to which societies thrive changes greatly depending on how production and distribution are organized and how social resources are shared. No matter how much the economy might grow, if the resulting wealth is monopolized by one part of the population and never redistributed, large numbers of people will live in comparative misery, unable to realize their potential.

This can be seen the other way as well: even if its economy doesn't grow, if existing resources are distributed well, a society may thrive more than ever. In the UN's *Human Development Index* (HDI), which calculates life expectancy, education, and per capita income, the US, despite being the world's greatest economic superpower, ranks only thirteenth. The highest-ranked country is Norway.

It's clear that the GDP is not the whole story. Rather, we must think more deeply about whether a just distribution of resources can occur on a consistent basis within a capitalist system.

TOWARD A JUST DISTRIBUTION
OF RESOURCES

The difficulty that arises at this juncture is that a just distribution of resources is not a problem to be solved at the national level. The enormous question here is how we can realize a just and sustainable society at the *global* level.

Please don't misunderstand me—I don't mean to be hypocritical here. The climate change problem demonstrates quite clearly that the entire world is connected. It's simply unsustainable for Global North countries to continue their wasteful consumption and demand that emerging nations follow the same path of development in order to sell them the goods they wish to keep consuming. If the entire world, without exception, does not become part of a sustainable, just society, the environment will deteriorate until the planet is unlivable for everyone, threatening even the prosperity of the Global North.

It's imperative that we bring those living below the level of Raworth's social foundation up into the range of sustainable prosperity. But this will inevitably contribute to an increase in the overall global material footprint. Such an increase, given the current state of affairs in which many places are already overshooting the planetary boundaries of sustainability, would be fatal to the environment.

Therefore, the countries of the Global North must not continue to use enormous amounts of energy to foster even more economic growth. As we have seen, increasing economic growth past the current level is hardly guaranteed to improve the level of happiness and well-being of the people living in those countries.

But if the same resources and energy were used in the Global South instead, the happiness and well-being of the people there would increase exponentially. And if this is true, doesn't it follow that we should set aside a part of the worldwide "carbon budget" (the amount of carbon dioxide emissions the world can still safely allow) for them?

In short, if we don't think, "What does it matter if the billion people currently starving on this planet continue to starve?" or

"Who cares if future generations are left to suffer in an uninhabitable environment?" we must seriously consider following a path away from the large-scale, wasteful cycle of production and consumption of the Global North and toward the scaling down of our material footprint, starting at home.

It's for this reason that both Raworth and O'Neill conclude their studies by saying that we should seriously consider transitioning to either a "degrowth" or "steady-state" economy.[59] I want to state clearly here that at least up to this point, I agree with their overall assessment.

CAPITALISM CAN NEVER BRING ABOUT GLOBAL JUSTICE

There is, however, a major feature of both Raworth's and O'Neill's arguments that leaves me in grave doubt: their shared unwillingness to question the capitalist system itself. This is where we can glimpse the distinctive characteristics of people pushing for degrowth while avoiding problematizing capitalism. The crux of the problem here is whether a truly just distribution of resources can ever be sustained under capitalism.

From a global justice point of view, capitalism is a completely dysfunctional, profoundly unhelpful system. As demonstrated in the previous two chapters, capitalism's dependence on externalization and displacement guarantees that it will always work against global justice. Such an abandonment of justice will only result in humanity's survival becoming less likely.

In short, it is insufficient to address the environmental crisis by striving to save only ourselves. We may buy a little time that way, but in the end, there is only one planet. Soon enough, there will truly be nowhere left to go to escape the crisis.

The lifestyles of those living in the top 10–20 percent of the wealthiest nations in the world seem, at present, still stable. But if we continue our current modes of living, the global environmental crisis will only worsen. Eventually, only the top 1 percent of the richest people in the world will be able to continue their present lifestyles.

For this reason, the issue of global justice is neither abstract nor a matter of utopian humanitarianism. Before giving up on helping others, take a moment and imagine yourself standing in their shoes tomorrow. In the end, the only way to help ourselves and improve the chances of survival for all of humanity is to help bring about a just, sustainable society for everyone.

The key to survival is, above all, equality.

FOUR CHOICES, FOUR FUTURES

What will the future look like as we make the choices necessary to survive the Anthropocene, now that we know that the key to this survival is equality? Let's pause here and take a broad look at our options.

The horizontal axis of Figure 6 shows levels of equality, starting at the left with the most egalitarian position and moving rightward toward one emphasizing total self-reliance. The vertical axis charts the concentration of political power, with state power increasing as one moves upward and mutual aid between individuals increasing as one moves down.

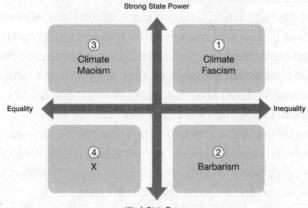

Figure 6. Four choices, four futures

Now, let's look at four possible outcomes for the future:[60]

1. *Climate fascism*

If we choose to do nothing and keep pursuing economic growth through capitalism in order to support the status quo, the damage brought about by climate change will become enormous. Before long, many people will no longer be able to sustain even the most basic lifestyle. Large numbers of people will lose their homes, becoming climate refugees.

However, things will be different for the superrich elite. Those who benefit from disaster capitalism will see the crisis as a business opportunity enabling them to become even richer. Nations will want to protect the interests of this special class of elites, and they will crack down on the climate refugees and other vulnerable populations that would threaten to overturn this order. This is the first future: climate fascism.

2. *Barbarism*

At the same time, if climate change continues to advance, climate refugees will proliferate and food production will collapse. As a result, populations suffering from starvation and poverty will start civil wars and uprisings. If there is truly a war between the top 1 percent and the remaining 99 percent, the latter will surely win. A rebellion of the masses would overturn authoritarian governments attempting to rule with iron fists, plunging the world into chaos. All confidence in the rule of law would be lost, and the world would descend into a Hobbesian state of nature, a "war of all against all" in which everyone would look out only for their own individual interests. This is the second future: barbarism.

3. *Climate Maoism*

Some form of rule would be necessary to save society from the second-worst-case scenario of a descent into barbarism. One way this might occur is by narrowing the wide wealth gap dividing the top 1 percent from everyone else and instituting top-down climate change policies. This would involve jettisoning the free market and liberal democracy and creating a centralized authoritarian dictatorship that would bring about "more efficient" and "equal" ways to combat climate change. This is the third future: climate Maoism.

4. X

There must be a fourth way, a way to resist both a slide into barbarism and some form of autocratic nationalism. It's not impossible to implement democratic forms of mutual aid that don't rely on a strong nation to function, instead relying on individuals to voluntarily develop strategies for combating climate change on their own. This would be a just, sustainable future society. This fourth future has no name; for now, let's call it X.

The future I am calling for in this book is, as you might guess, the fourth one. Out of all the possible futures, it's the only one offering a chance at survival while preserving ideals of freedom, equality, and democracy. The aim of the rest of this book will be to elucidate the exact nature of this unknown future and how to achieve it.

WHY CAN'T WE ACHIEVE DEGROWTH UNDER CAPITALISM?

The thing is, it's not true that we have no idea how to bring about X. I've been hinting at it for several pages already. The key is degrowth.

Why is degrowth necessary to fix the climate crisis? My sense is that what we have seen so far is answer enough. We saw how "green economic growth" fails to lead to the preservation of a global environment capable of supporting all of humanity. Sufficient absolute decoupling is nothing but an illusion, and while the "green" label may be trendy, even "green" growth will inevitably lead to insupportable increases in the burden on the environment. Government policies intended to foster economic growth will never bring about an end to the global environmental crisis.

Therefore, we need a new logic, one separate from that of so-called green Keynesianism. Degrowth is a choice we must make to create an economic system that no longer depends on growth to function. This is the conclusion of scholars like Raworth. Degrowth is meant to put the brakes on capitalism run amok and bring about a type of economy that would prioritize the needs of both humanity and nature. So far so good. I am in complete agreement. But is there really a form of degrowth that can be sustained within the capitalist system? This is the question we must seriously consider at this point.

As I will demonstrate, the solution to this problem cannot be the tepid call to modify neoliberalism and tame the capitalist system until degrowth can be brought about within it, as is argued in the work of Raworth and others. This is because the devil destroying the global environment is none other than the capitalist system itself and its demands for constant unlimited growth. It's capitalism—nothing more, nothing less—that lies at the root of climate change and the other global environmental crises that come with it.

Capitalism is a system that always seeks to open up new markets to accumulate capital and increase value. This process necessitates the exploitation of natural resources and human labor, a cost that's constantly externalized onto the periphery. As Marx put it, this is a movement with "no limits." The essence of capitalism is that it will never stop growing in order to create profit.

This is why at this juncture, we can no longer afford to choose capitalism. Capitalism sees even the worsening of the global environmental crisis, climate change and all, as just another opportunity to make money. A rampant wildfire is an opportunity to sell wildfire insurance; a plague of locusts is an occasion to sell more fertilizer. So-called Negative Emissions Technologies, as we've learned, may produce side effects that ruin the planet, but even these can be seen as yet more business opportunities. This is precisely what Naomi Klein characterized as "disaster capitalism."

Even as the number of people suffering goes up as the crisis worsens, capitalism will continue to seek opportunities to profit until the very last moment, exhibiting its characteristic tenacity by adapting to any condition. It will never stop itself, even when faced with something as terrifying as the environmental crisis.

This is why, if we allow things to continue as they are, capitalism will transform every inch of the Earth's surface into an environment unable to support human life. This is the endpoint of the Anthropocene.

As mentioned in the previous chapter, there are some who say that one potential way to head off the climate crisis would be to live as people did in the late 1970s. Hearing that, one might well reply that the 1970s were a capitalistic era as well, so why wouldn't a return to 1970s capitalism be a way to avoid the crisis?

Let's remember that capitalism was experiencing a profound system failure during the 1970s. To pull out of this tailspin, the political program now known as neoliberalism was introduced. Neoliberalism promoted privatization, deregulation, and austerity while expanding financial markets and free trade, setting the world on the path to globalization. This was the only way for capitalism to survive.[61]

Besides, even if it were possible to return to 1970s capitalism (and it's not), capitalism, which always aims to increase capital, would hardly be content to stay at that level. If it were to do so—if it were to abandon profit making—it would risk falling into the same system failure that threatened it in the real 1970s. It would end up forging the same path as it did before to save itself, and in the end, the environmental crisis would just worsen once again.

Therefore, the only real way to face the environmental crisis head-on and control economic growth is to give up capitalism and bring about a great transition to a post-capitalist system of degrowth.

WHY DOES POVERTY CONTINUE?

Even as I argue that degrowth is the option we must choose to overcome the crisis, I can imagine many readers will deny its necessity and argue against it.

One reason may be that the word "degrowth" sounds to some like another word for "voluntary poverty," and the only people singing the praises of downsizing and letting go of material possessions are those rich enough to be blissfully unaware of the plight of the average worker. If growth stops at the macro level and the pieces of pie available for redistribution never get bigger, wealth will never reach the poor—in short, nothing will be able to trickle down.

On the surface, this criticism is correct. The present system is based on growth. In such a society, if growth stops, it would of course lead to tragedy. But the question remains: In a world in which capitalism has progressed as far as it has, isn't it strange that so many living even in Global North countries continue to languish in poverty, their lives only getting harder as time goes on?

Our salaries seem to disappear before we know it, consumed by commuting and paying rent, phone bills, bar tabs, and the like. We have to scrimp and scrape, always cutting back on what we spend on food, clothing, and a social life. And then, as we barely manage to save what we need to support ourselves, we end up loaded down with student loans and mortgages, and we work as hard and long as we can every single day. If that's not a form of voluntary poverty, then what is?

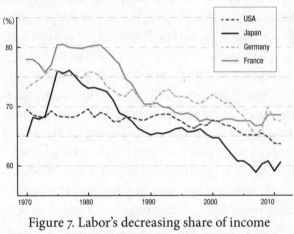

Figure 7. Labor's decreasing share of income
Based on OECD

How much more economic growth has to take place before we get rich? Won't the path of continued growth just lead to more painful restructuring and quantitative easing as the share of income going to labor decreases and the gap between the rich and poor widens? And on top of all that, how much longer will we be able to sacrifice our natural world in the name of fostering growth?

GENERATION Z'S CRITICISMS OF CAPITALISM

Even as the call to continue economic growth sounds more and more illogical, degrowth remains unpopular, due in part to a generational problem. There is a strong tendency to imagine that those supporting degrowth are simply paying lip service to a nice-sounding idea, that they are part of a generation who reaped the benefits of high-speed economic growth but want to avoid its consequences. Having enjoyed the fruits of growth in their youth, they're now content to shut the door behind them and say, "What does it matter if the developed world slowly declines?" This forms much of the impetus behind the criticism of the idea coming from the younger generation.

This sort of Baby Boomer–style argument for degrowth has produced, as its antithesis, growing support for extreme "anti-austerity" positions like Modern Monetary Theory, a heterodox economic theory according to which the government, as the monopoly issuers of currency, can print as much money as they like without causing inflation. Of course, anti-austerity that prioritizes people's quality of life is a wonderful thing. As touched on in chapter 1, when we talk about politicians and economists calling for the end of austerity, the centerpiece of their anti-austerity measures is a Green New Deal. In other words, they promise to revolutionize modes of production and social infrastructure as a way to combat climate change.

The "leftist populism" of figures like Bernie Sanders is largely supported by young millennials and members of Generation Z. The distinctive features of these generations include very high levels of concern for the environment and skepticism about capitalism. Indeed, some have called them Generation Left. Some public opinion polls even show that more than half of the American members of Generation Z prefer positions aligned with socialism over capitalism.[62]

As is often pointed out, Generation Z—which includes those born between 1990 and the 2000s—are digital natives and have always been able to use the newest technologies to connect with friends around the world. In this sense, they have strong identities as global citizens.

Above all, though, this young generation has experienced first-hand the widening of the income gap and the destruction of the environment that has accompanied the financial deregulation and privatization promoted by neoliberalism. There seems to be no bright future waiting at the end of capitalism's continued advance, only the ongoing need to clean up the messes left by adults. What can they do in the face of this but despair and grow angry?

This is why members of Generation Z, conscious of themselves as citizens of the world, are currently trying to change that world. Greta Thunberg is in this sense a representative member of this generation.[63]

THE GREEN NEW DEAL AS COMPROMISE

But will anti-austerity measures like the Green New Deal proposed by Bernie Sanders and his ilk really be enough for a generation gripped more and more by fear of the impending crisis? Here two problems present themselves.

The first is that while the Green New Deal proposed by leftists naturally takes a critical stance toward neoliberalism, the theoretical foundations of the deal, as theorists like Robert Pollin and Noam Chomsky have pointed out, amount to nothing more than reforms conducted within a capitalist framework. As long as this is the case, what will prevent the Green New Deal from flirting with the kind of business-friendly "green growth" policies trumpeted by Thomas Friedman and the like until it becomes indistinguishable from them?

The second problem is the one we examined in the previous chapter: Can a Green New Deal conducted under capitalism solve the problem in time? As long as reform takes place within a capitalist framework, it will always tend to privilege growth. As we've seen, this will inevitably lead to an inability to realize sufficient absolute decoupling, amounting to nothing more than an ineffective half measure.

In other words, anti-austerity is a form of compromise. Even as it critiques neoliberalism, it insufficiently critiques capitalism. After all is said and done, it's clear these programs will never truly foster degrowth. What's truly necessary to address the needs of the younger generation is a Green New Deal without Growth.[64]

DEGROWTH-CURIOUS

Older generations think of degrowth as a politically impossible choice. As long as economic growth continues and its blessings are shared among the majority, people will be satisfied and society will be stable. However, if this growth is obstructed, the economic gap will widen and environmental problems will get worse. This is what the Anthropocene is. The result is that the validity of the way we have done things up until now is thrown into question.

This is why there are "revolutionary" environmental movements rising up all over the world that favor direct action. Groups like Extinction Rebellion in the UK and the Sunrise Movement in the US are developing modes of resistance that involve direct actions like strikes, sit-ins, and property destruction, sometimes performed even in the face of certain arrest. They are gathering supporters from all walks of life, including not only average citizens and students but Hollywood actors and Olympic gold medalists as well. Their voices are shaking the foundations of the ruling class's credibility, paving the way for a new set of political possibilities. These are movements that have the potential to overcome capitalism as well.

If, by contrast, the liberal left continues to simply turn its back on the worsening state of the environment and keep calling for economic growth, then its anti-austerity measures will only go as far as green Keynesianism is willing to go, and all their efforts will wind up being just more ways for capitalism to restabilize itself.

The present age of environmental crisis offers us a chance to step through the door leading to a much more revolutionary, complete political transformation than that. Yet these voices offer more of the same calls to foster the very economic growth that caused the destruction of the environment in the first place rather than unleashing the power to imagine a society that's truly different.

If we resign ourselves to such a limited imagination of political possibility, carbon dioxide emissions will simply continue to increase. And we will face the justified condemnation of Generation Left all the more.

THE LIMITS OF THE DEGROWTH IMAGINED BY PREVIOUS GENERATIONS

In truth, the concept of degrowth emerged back in the 1970s, and wider support for it started to accumulate in the 1990s. This book is heavily influenced by the ideas of one of the central figures in this scene, André Gorz. But his theory of degrowth is already outdated.

What's outdated about it? Older degrowth theory seems, on the surface, critical of capitalism, but in the final analysis, it ends up accepting it. Whenever we attempt to argue for degrowth within the framework of capitalism, we inevitably find ourselves mired in negative images like "stagnation" and "decline."

This limitation can be traced to the historical background against which older theories of degrowth were formulated, namely the fall of the Soviet Union. As the world-famous French first-generation degrowth theorist Serge Latouche put it, they see Marxism as nothing more than the fantasy of an "impossible return to the past."[65] In this context, degrowth becomes merely an attempt to reinstate liberal leftism as the order of the day.

The outdated degrowth project put forth by Latouche and others belongs properly to neither the left nor the right. This is because they see nature as a matter of universal concern, whether one happens to be rich or poor, a leftist or a rightist. For this reason, the previous generation of degrowth theorists weren't aiming to move past capitalism. Instead, they fundamentally rejected it as a framework for understanding the issue.

JAPAN'S DEGROWTH OPTIMISTS

Degrowth theories of previous generations that are unwilling to aim for the end of capitalism exist where I live, in Japan, as well. For example, Professor Yoshinori Hiroi, who has been influential in promoting the concept of a steady-state society in Japan, defines his proposed society as one featuring a "sustainable welfare state and welfare society." As he puts it:

> One thing I would like to make clear right from the outset is that the steady-state society I am envisioning would not involve the abandonment of a market economy or the pursuit of self-interest. In other words, a steady-state economy is *not* a socialist economic system . . . it's a conceptualization of society that transcends received binaries of capitalism versus socialism or freedom versus equality.[66]

Furthermore, socioeconomic theorist Keishi Saeki, after dismissing the idea that he is promoting "the escapist path known as socialism," defines his project thusly:

> In the present environment of competition-based economies and competitive growth, the promotion of accelerated growth at any cost by monetary authorities all over the world via the supply of excessive amounts of liquidity will lead to the destabilization of money markets, including the creation of bubbles that will inevitably pop . . . Degrowth is nearly the only way left to insure that capitalism remains stable.[67]

According to Hiroi and Saeki, capital growth can be halted while supporting a capitalistic market economy. The problem is that capitalism has indeed gone too far, but after the collapse of the Soviet Union, there can be no return to socialism. Welfare state policies characteristic of social democracy are presented as ways to tame neoliberal market fundamentalism. It is at this juncture that ideas of sustainability are introduced. This is how they envision the possibility of transitioning to a degrowth or steady-state society.

If this is true, there is no need to bring about any fundamental changes to the relationships between wage labor, capital, and private property, nor to the competition for profits in the marketplace. All that needs to be done is instituting appropriate incentives and organizational planning to fully mature nations as their material consumption reaches a state of saturation. As long as this occurs, the argument goes, people will voluntarily engage in the various activities associated with the social and public sphere that fall outside the pursuit of profit in the free market.

THE STARTING POINT FOR A NEW THEORY OF DEGROWTH

But isn't such a hypothesis overly optimistic? This question is the starting point for a new theory of degrowth. It's true that a return to a Soviet-style arrangement is out of the question. But the new degrowth theory states that any attempt to blend degrowth with capitalism is doomed to fail, so capitalism must be challenged as well.

To clarify what this might mean, I want to introduce an argument made by the Slovenian Marxist philosopher Slavoj Žižek in response to Nobel Prize–winning economist Joseph E. Stiglitz, which can double as a cogent critique of older degrowth theory as well.

Stiglitz became famous for his harsh criticism of the excesses of globalization, the uneven distribution of wealth in the present day, and the domination of the market by large corporations. The object of Žižek's criticism, though, is the solution Stiglitz proposes for these issues, which the latter calls "progressive capitalism."

Stiglitz rejects faith in the free market, stating that raising workers' wages and additional taxation of the superrich and major industry, along with stricter prohibitions on monopolies, are necessary to bring about a just capitalist society.[68] If democratic elections were to change laws and policies along these lines, then economic growth could be revived, and a "progressive" capitalism in which the average person would belong to a comfortable middle class might be possible.

Can capitalism really be tamed by changing laws and policies? This is the crux of Žižek's skepticism. If increases in taxation, including corporate taxes, and social welfare spending were possible to bring about via elections, why hasn't it already happened? When profit rates in the 1970s fell, plunging capitalism into a very serious crisis, various regulations were quickly jettisoned and taxes were lowered right away. If regulations were put back in place at the same level as they were then or even a higher level, wouldn't capitalism just collapse again? In any case, it would never accept this sort of restriction on its functioning—it would oppose it at every turn. Capitalists have the power to exert enormous amounts of political pressure, not just by threatening to move factories overseas but to sell government bonds and cause stock prices to fall precipitously.

Stiglitz, in short, poses his vision for a just future as "true capitalism" and the present arrangement as "phony capitalism," but this overlooks the following possibility: that the "golden age" of capitalism—from the end of World War II until the 1970s—to which he so longs to return was in fact the aberration, the "phony" version of capitalism. That would mean the present "phony" system that Stiglitz so abhors is, in fact, capitalism's true face.

In this sense, the reforms Stiglitz calls for are incompatible with the continued workings of capitalism, and for that reason they can never actually be realized. Doesn't this make Stiglitz, who passionately argues that these reforms are exactly what will sustain capitalism into the future, the one chasing dreams of an impossible return to the past?[69]

DEGROWTH CAPITALISM CAN NEVER EXIST

Thinkers who propose transitioning to a degrowth society while preserving capitalism are the ones who truly deserve to be called fantasists. Capitalism's very definition precludes any possible pairing with degrowth.

Capital is defined by an endless movement to increase value. Investment is repeated again and again, while the production of goods and services creates new value, raising profits and continuously expanding. The only way it can realize its goals is to use all the world's resources and labor power, opening new markets and never passing up even the smallest chance to make more money.

But the expansion of capitalism across the surface of the entire Earth has resulted in the destruction of both the natural world and humanity's quality of life. For this reason, degrowth is necessary to put the brakes on capital and decelerate its movement.

The older generation of degrowth theorists argue the following: that capitalism must stop externalizing its contradictions and transferring the costs to the periphery. It must also stop its plunder of natural resources. We should stop prioritizing corporate profits and instead place our emphasis on enhancing the happiness of workers and consumers. The scope of the market must also contract to a level that's sustainable.

This indeed would be a carefree form of "degrowth capitalism." But the problem is, profit making, market expansion, externalization, displacement, and the extraction of human labor and natural resources are what constitute capitalism's very essence. To demand the cessation of all these things—to demand deceleration—is in fact to demand capitalism's end.

In other words, the desire to do away with the urge to grow in order to maximize profits—capitalism's most fundamental characteristic—while also preserving capitalism is like trying to draw a round triangle. It's nothing more than a fantasy. This is the true limit of the previous generation's conceptualization of degrowth.

WERE JAPAN'S "LOST DECADES" DEGROWTH?

Let's think about the impossibility of effecting degrowth under capitalism in more detail, using Japan as an example.

The situation in Japan during the three "lost decades" following the bursting of the bubble economy in 1991 was, in truth, a prime example of degrowth occurring while capitalist growth imperatives still operated. In fact, the book *Slowdown* by Oxford professor Danny Dorling uses Japan as an example of a mature society in a position to prioritize a new form of "good life."[70]

But there is no worse situation for capitalism than one in which growth is no longer possible. When growth stalls under capitalism, industry strives even more desperately to raise profits. In this zero-sum game, workers' wages go down while restructuring and the elimination of permanent jobs proliferate as cost-cutting measures. The class divide widens within the country while exploitation only intensifies in the Global South.

The share of wealth going to labor has indeed gone down, and the gap between the rich and poor has steadily widened. In Japan, so-called *burakku kigyō* (literally, black companies) chewing through young workers by subjecting them to inhumane conditions has become a serious social problem. As the pie gets smaller and stable employment becomes scarce, people compete ever more fiercely with each other for survival. In Japan, the recent popularization of referring to people as high-class/low-class citizens indicates how deeply this social divide has been inscribed into people's hearts.

REEXAMINING THE MEANING OF DEGROWTH

The unfortunate state of Japan shows us something important. Phenomena like Japan's long-term stagnation or the global recession caused by COVID-19 must not be confused with a steady-state or degrowth economy.

It's a common misconception that the prime objective of degrowth is reducing the GDP. This leads to the GDP becoming the only figure people look at in the conversation about degrowth.

Under capitalism, we have consistently sought to raise the GDP in the belief that economic growth brings prosperity to everyone. But this prosperity has yet to arrive for the average person. In truth, the GDP is an extremely superficial indicator developed around a hundred years ago and one that has enormous statistical limitations. Given how much we've progressed since then, why are we still allowing ourselves to be manipulated by this crude measure?

As capitalism's antithesis, degrowth emphasizes forms of prosperity and quality of life that aren't necessarily reflected in the GDP. Degrowth is a transition from quantity (growth) to quality (flourishing). It's a grand plan to transform the economy to a model that prioritizes the shrinking of the economic gap, the expansion of social security, and the maximization of free time, all while respecting planetary boundaries.

If new coal-burning power plants continue to be built, as they do in Japan, degrowth is not taking place. If growth is stalling but the gap between the rich and poor is still widening, degrowth is not taking place. Even if production shrinks, the resulting rise in unemployment is a far cry from "maximizing free time." What needs to be reduced is the number of SUVs and the amount of beef and fast fashion being consumed, not funding for education, social security, and the arts.

In short, unlike what Hiroi asserts, Japan is far from being in a "leadership position" regarding degrowth. All that's happening is capitalism's long-term stagnation.

TOWARD A FREE, EQUAL, AND JUST THEORY OF DEGROWTH!

Degrowth aims to bring about equality and sustainability. By contrast, the long-term stagnation of capitalism brings about nothing but inequality and want, which then leads to intensified competition between individuals.

In today's society, which exposes everyone to relentless competition, no one has the luxury to extend a hand to those less fortunate. If you become homeless, you can even find yourself turned away from an evacuation center during a hurricane. In a dog-eat-dog society wherein a lack of money can lead to the deprivation of basic human rights and even threats to one's life, mutual aid becomes nearly impossible.

Therefore, if we're serious about wanting to foster mutual aid and equality, we must seriously reconsider issues of class, money, and market. A true transition to a degrowth or steady-state economy cannot be brought about by laws and policies meant to prioritize sustainability and the redistribution of resources as long as the fundamental essence of capitalism is left intact.

Even Kate Raworth stops before going that far. The key sectors of the doughnut economy she envisions include population, distribution, technology, aspiration, and governance.[71] What she fails to mention as fundamental issues are production, market, and class—in other words, the capitalist mode of production.

Can enough of a brake be put on capitalism to make it sustainable without touching on issues of private property and class? This sort of position amounts to nothing more than a capitulation to the power of capital and a way to guarantee the continuation of the inequality and lack of freedom promoted by capitalism.

In the end, the prospect of degrowth capitalism sounds attractive but turns out to be impossible to realize—just another fantasy. It appears in none of the four futures previously outlined (see Figure 7). The future known as X, the one I am arguing for, is not degrowth capitalism.

If we wish to champion degrowth, we must grapple with a more difficult theoretical and practical problem, one that cannot be solved by blending degrowth with capitalism. It's imperative that we face the present-day Great Divergence firm in our resolve to challenge capitalism itself.

We must bring about a free, equal, just, and sustainable society that overcomes class divides of exploitation and domination and that radically revolutionizes labor. This is the true face of degrowth for a new generation.

RESURRECTING MARX IN THE ANTHROPOCENE

Looking back over the course of history, how can we sincerely believe that mature capitalism will accept low or no growth and "naturally" transition to a steady-state economy? Historical precedent tells us that what really awaits us during a period of low growth is an intensification of ecological imperialism and climate fascism tied to maintaining the Imperial Mode of Living.

Along with this will come disaster capitalism, which aims to take advantage of the chaos engendered by the climate crisis. But if we continue on this course, the global environment will worsen to the point that it can no longer be remedied by human hands, and society will descend into a state of barbarism. This is the hard landing of slowed growth. It is, of course, another scenario to be avoided at all costs.

To head off the hard landing of the Anthropocene, we must have a theory—and a practice—that is willing to criticize capitalism in plain terms while demanding the active transition to a degrowth society equally plainly. There is no time left to waste on half measures or policies that would defer action to some future date. The degrowth theory of the new generation must incorporate a much more radical critique of capitalism to work. Yes, I am talking about communism. Hence the necessity of bringing degrowth together with the writings of Karl Marx.

There are surely many readers who will resist not just the evocation of Marx but any attempt to use his ideas to theorize degrowth. After all, isn't Marxism a theory of class conflict with nothing to contribute to a conversation about the environment? And isn't it the case that the Soviet Union was so preoccupied with economic growth that it did significant amounts of environmental damage? Isn't combining Marxism and degrowth like trying to mix oil and water?

The truth, as I will show in the next chapter, is quite different.

Let us call on Marx to awaken from his long slumber. He will surely heed our cry, even as it arrives all the way from the Anthropocene.

4

Marx in the Anthropocene

REHABILITATING MARX

In the face of the environmental crisis of the Anthropocene, we've seen that there is no alternative to toppling capitalism and imagining a postcapitalist future for ourselves. Even so, why should we, at this late date, return to Marx?

A common image people have of Marxism involves the nationalization of modes of production accompanied by one-party rule in the style of the Communist Parties of the Soviet Union or the People's Republic of China. With this in mind, there are surely many readers who see this political ideology as both outdated and dangerous.

It's true that the collapse of the Soviet Union led to a general stagnation in interest in Marxism. Even among those who identify as leftists, there are very few who openly champion Marx and advocate for the application of his ideas.

However, there are recent signs that this is changing slowly, and Marx's ideas are receiving increasingly widespread attention. As the contradictions of capitalism have deepened under neoliberal reforms to increase economic inequality, low-paid precarious jobs, and financial cuts in the social welfare system, cracks have started to appear in the formerly rock-solid commonsense assumption that "there is no alternative." This tendency is accelerating, as younger generations who were not alive during the Cold War and the collapse of the USSR are more open-minded about Marxist ideas. As we saw in the previous chapter, some polls have even shown that younger generations in the US view socialism more favorably than capitalism.

This chapter aims to clarify how Marx's ideas might apply to the environmental crisis of the Anthropocene and what we might glean from this as we attempt to formulate solutions to the crisis that differ from those offered by green Keynesianism.

This will not be a rehash of outdated Marxist analyses. Rather, I intend to use fresh archival sources to present a new Marx for a new era: the Anthropocene.

A THIRD WAY—THE COMMONS

A key concept that has arisen in the recent reevaluation of Marx is the idea of "the commons." The commons is a term for forms of wealth that should be managed and shared by every member of a society. It's a concept that gained renewed popularity after being featured in *Empire*, a monumental book by the Marxist scholars Michael Hardt and Antonio Negri that was published in English in the year 2000.[72]

The commons may be thought of as the key to a third way that would represent an alternative to the opposing extremes symbolized by US-style neoliberalism and Soviet-style nationalization. It can be understood as a way to avoid commodifying necessities according to the principles of market fundamentalism while stopping short of a Soviet-style nationalization of properties. The commons, a third way between these extremes, aims to designate things like water, electricity, shelter, healthcare, and education as public goods and manage them democratically.

We can also think about this using the concept of "social common capital" as advanced by the economist Hirofumi Uzawa.[73] According to Uzawa, certain basic conditions must be satisfied for people to thrive in a "rich society." These conditions include aspects of the natural environment such as water and land, social infrastructure like electricity and public transportation, and social systems like healthcare and education. These things should be thought of as common goods for all of society and thus should be managed and operated socialistically, exempt from market norms and national regulations. It's an idea that's basically identical to that of the commons.

The main difference is one of emphasis, with the commons prioritizing shared management by citizens in a democratic, equal way rather than leaving administration up to specialists, as advocated by the concept of social common capital. The other decisive difference is my aim to gradually expand the commons until, in the end, they displace capitalism entirely.

MANAGING THE COMMONS KNOWN AS THE PLANET

In fact, Marx's version of communism did not aspire to Soviet-style one-party rule and nationalization. Rather, communism was, for Marx, a way to bring about a society in which producers shared the means of production, managing and operating them together as a form of commons.

Marx states as much in a famous line from near the end of *Capital*'s first volume. He describes the advent of communism as when "the expropriators are expropriated," the natural outgrowth of the "negation of negation" of capitalist production:

> [The negation of the negation] does not re-establish private property, but it does indeed establish individual property on the basis of the achievements of the capitalist era: namely co-operation and the possession in common of the land and the means of production produced by labor itself.[74]

What does Marx mean by the "negation of negation"? Let me provide a brief explanation. The first negation occurs when producers find themselves cut off from the means of production formerly held in common and must instead work for capitalists. Marx calls this the primitive accumulation of capital. The second negation refers to when workers release the property held by capitalists. This means that the land—that is, the Earth—and the means of production are returned to the commons.

Of course, this formulation remains an abstract schema. But Marx's views here are clear. Communism is meant to topple the capitalism that destroys the Earth in its pursuit of endlessly increasing profits. We must all manage the Earth together as the ultimate commons.

RECONSTRUCTING THE COMMONS THROUGH COMMUNISM

Emphasizing the commons as a fundamental part of Marx's thought is not confined to theorists like Hardt and Negri. Žižek, too, calls for the necessity of communism as he examines the idea of the commons.

According to Žižek, there are four types of commons: the commons of culture, the commons of external nature, the commons of internal nature, and the commons of humanity itself. Global capitalism advances by the "enclosure" of these commons as antagonisms dividing the populace. He states, "It is this reference to 'commons' which allows the resuscitation of the notion of communism."[75]

As Žižek says here, communism is nothing less than the conscious attempt to reconstruct the commons—knowledge, nature, human rights, society—dismantled by capitalism.

It's not well known, but Marx in fact referred to the society to come, a society founded on the reconstructed commons, as one of "free association." When he spoke of this future society, Marx seldom used the words "socialism" or "communism." Rather, he preferred to refer to the "free association" of producers. The voluntary mutual aid between workers characterizing such an association is the ultimate realization of the commons.

SOCIAL SECURITY BORN OF FREE ASSOCIATION

This meaning of the term "commons" doesn't refer to a new need that arrived with the advent of the twenty-first century. The social security services now provided by the state were once commons that originally arose from free associations between people.

In other words, social security originated as part of a series of efforts people made to provide each other with things necessary to live good lives without relying on the market. What happened in the twentieth century was the systematization of these efforts by the welfare state.

London School of Economics professor and cultural anthropologist David Graeber explains it this way:

> In Europe, most of the key institutions of what later became the welfare state—everything from social insurance and pensions to public libraries and public health clinics—were not originally created by governments at all, but by trade unions, neighborhood associations, cooperatives, and working-class parties and organizations of one kind or another. Many of these were engaged in a self-conscious revolutionary project of "building a new society in the shell of the old," of gradually creating Socialist institutions from below.[76]

Graeber is explaining how the commons formed out of free association were systematized under capitalism into the welfare state. But starting at the end of the 1980s, neoliberal austerity measures dismantled or weakened associations like labor unions and public health services one after another, allowing these commons to be swallowed by the market.

Resisting neoliberalism and returning to the welfare state model is, unfortunately, an insufficient countermeasure at this juncture. The road represented by the welfare state, based as it is on high-speed economic growth and the North-South Divide, cannot lead us into a sustainable future in the age of climate crisis, as the best-case scenario is a descent once more into a nationalistic version of green Keynesianism. There is also the danger that this could easily slip into something like climate fascism.

The nation-state form is completely unable to adequately address the present global environmental crisis. Furthermore, the vertical nationalized management style characteristic of a welfare state is incompatible with the horizontal nature of the commons. In other words, the commons must be made sustainable at a global level, not turned into something to enrich the lifestyles of some. To this end, they must be reappropriated from their commodification by capital. There must be another way to address the problem; a larger vision is necessary. Only an unprecedented form of Marxist analysis can answer the demands of the era of environmental crisis known as the Anthropocene.

A NEW COLLECTION OF THE COMPLETE WORKS OF MARX AND ENGELS—THE MEGA PROJECT

Some may very well doubt that a new Marxist analysis could ever arise now, in the twenty-first century. Wouldn't any such attempt amount to nothing more than old ideas dressed up in new clothes? And indeed, there are many books out there that are just that.

But in this case, there's a brand-new publication project under-way called MEGA—*The Marx–Engels Gesamtausgabe* (*The Complete Works of Marx and Engels*). It's an international project that includes scholars from all over the world, including me. The scope of the proj-ect is unprecedently huge and may well end up numbering more than one hundred volumes. It includes many new materials that have never been made public before.

Among these new materials, Marx's "Research Notes" deserve special attention for the insights they provide. Marx had a habit of writing out long excerpts from sources in his notes as he conducted research. Lacking funds while he lived in exile, he would go to the

British Library in London every day and borrow books, writing excerpts into his notes as he looked through them in the reading room. The notes he amassed while doing this are extensive and include ideas and complexities that never made it into *Capital*. For this reason, they are precious primary sources.

Up until now, these notes were dismissed as mere excerpts from other sources, neglected by scholars and never published. Now, though, they are being collected for publication as the thirty-second volume of the MEGA project's fourth part.

What the MEGA project has thus made possible is a totally new interpretation of *Capital*, one that differs greatly from the general understanding of it. Scholars painstakingly deciphering Marx's near-illegible handwriting in the notes he left behind has resulted in new light being shed on *Capital*, illuminating a new weapon to use in the battle against the present climate crisis.

EARLY MARX AS PRODUCTIVIST

First, however, we should probably look at how Marx has been generally viewed up until now. The widespread understanding of his work goes something like this:

The advance of capitalism involves the extreme exploitation of workers by capitalists, widening the gap between the rich and poor. Capitalists compete against each other, raising their productivity, which leads to the production of more and more commodities. But workers, exploited due to low wages, can't afford to buy those commodities. This eventually leads to a crisis of overproduction. The already-exploited workers thus suffer another blow, this time due to the unemployment stemming from this crisis, and rise up en masse, bringing about a socialist revolution. The capitalists are purged, the workers freed.

This is a very simplified, broad summary of the *Communist Manifesto* coauthored by Marx and Engels in 1848. Still young at the time, Marx embraced the optimistic view that capitalism could eventually be overcome by a socialist revolution sparked by economic crisis. The advance of capitalism would prepare the ground for revolution by raising productivity levels and creating a crisis of overproduction. For this reason, he thought that productivity would have to rise under capitalism to bring about socialism. This is what is known as productivism.

However, the revolution of 1848 ended in failure. Capitalism caught its breath and rallied. The same thing happened after the crisis of 1857. Faced with the tenacity of capitalism as it overcame repeated crises, Marx began to revise his assumptions. These new understandings can be found in *Capital*, which he published twenty years after the *Communist Manifesto*, and other writings that followed its publication. As easy to understand as the *Communist Manifesto* may be, it's insufficient as a basis for understanding Marx's ideas overall.

THE UNFINISHED *CAPITAL* AND THE MAJOR SHIFT OF MARX'S LATER THOUGHT

Scholars, naturally, have studied Marx's *Capital* extensively. What makes the situation complex is that Marx couldn't express his final conclusions to their fullest extent even in *Capital*.

This is because while the first volume of *Capital* was completed by Marx himself and published in 1867, the second and third volumes were left unfinished. The version of these volumes that we read today were edited and published by his comrade Friedrich Engels after Marx's death. As a result, there are several instances where Marx's later ideas ended up distorted because of overediting or the differences between Engels's views and Marx's.

Marx's criticism of capitalism, in fact, continued to deepen during the course of his strenuous efforts to complete and publish the next volumes of *Capital* after the publication of the first. Indeed, this ended up representing a major theoretical shift in his thought. If we want to survive the environmental crisis of the Anthropocene, it is this later development in Marx's ideology that we should attend to most urgently.

The problem is that this major shift is difficult to detect in the version of *Capital* currently in circulation. Engels sought to emphasize the systematic nature of *Capital* and ended up concealing the parts of it that were unfinished. In other words, the parts of these volumes in which Marx grappled theoretically with his analysis were removed from view, as was the very evidence of this struggle. As a result, only the handful of researchers who have studied the research notes Marx kept at the end of his life are aware of this shift in his thinking. This means that even specialized scholars and committed Marxists labor under a misunderstanding of Marx's ultimate views.

This misunderstanding has had major consequences; it wouldn't be hyperbolic to say that this distortion of Marx's thought resulted in the birth of the monster known as Stalinism and humanity's ongoing inability to look at the present environmental crisis directly in its hideous face. We must correct this misunderstanding before it's too late.

THE DISTINCTIVE FEATURES OF THE PROGRESSIVE VIEW OF HISTORY: PRODUCTIVISM AND EUROCENTRISM

What is this misunderstanding, then? To put it simply, it's the attribution to Marx of the optimistic idea that the modernization brought about by capitalism will, in the end, bring about the liberation of humanity. This is an idea that appears in its most classic form in the *Communist Manifesto*.

And indeed, this is how Marx was thinking at the time he cowrote the *Manifesto*. Capitalism, he thought, might bring about the exploitation of workers and the destruction of the environment for a time. But it would also bring about innovation through competition that would raise productivity. This rise in productivity would prepare the conditions for everyone in the future society to enjoy a rich, free lifestyle.

Let's call this way of thinking the progressive view of history. According to the most widespread understanding of Marx, he was a classic proponent of seeing history as progress. Furthermore, this view displays two distinctive features: productivism and Eurocentrism.

The first of these, productivism, refers to a form of celebratory modernization theory that states that as the productivity fostered by capitalism rises, the problems of poverty and environmental damage will naturally be solved, eventually leading to the liberation of humankind. This is a linear view of history as progress. According to such a view of history, Western Europe, with its high levels of productivity, has reached a higher step in history. This means that the rest of the world must be modernized under capitalism to catch up, just as Western Europe was. This is where Eurocentrism comes in.

Productivism and Eurocentrism are intimately linked to the linear view of history as progress. This progressive view of history—under the name "historical materialism"—has been rightly showered with criticism.

THE PROBLEM WITH PRODUCTIVISM

First of all, adopting the productivist position enables us to completely ignore the destructive effect production has on the environment. Productivism states that the liberation of humankind will arrive along with the completion of humanity's conquest of nature. Productivism overlooks the ugly truth that the rise in productivity under capitalism is the very thing destroying the environment.

It's the perceived link between Marxism and productivism that led to the former's repudiation by environmental movements in the second half of the twentieth century. And indeed, one reason for this criticism comes from Marx himself. As he explains in this famous paragraph of the *Communist Manifesto*:

The bourgeoisie, during its rule of scarce one hundred years, has created more massive and more colossal productive forces than have all preceding generations together. Subjection of Nature's forces to man, machinery, application of chemistry to industry and agriculture, steam-navigation, railways, electric telegraphs, clearing of whole continents for cultivation, canalization of rivers, whole populations conjured out of the ground—what earlier century had even a presentiment that such productive forces slumbered in the lap of social labor?[77]

Looking at just this paragraph, the critique of Marx as productivist is quite valid. Marx is rather naïvely celebrating productivity under capitalism, leading readers to assume that he thinks this productive force is what will create an affluent society and prepare the way for the liberation of the proletariat.

The notion that the creation of a future free society depends on a productive force that allows humanity to conquer nature leads to the view that nature's limits simply need to be overcome. This would imply that Marx's thought lacks an ecological dimension, that "red" and "green" can never mix. This is one reason why Marxism has fallen out of favor in recent years.

THE BIRTH OF MARX'S THEORY OF METABOLISM—THE ECOLOGICAL SHIFT IN *CAPITAL*

But we've seen that this is not the whole story. We saw how deeply and incisively Marx examined the relationship between capital and the environment. We've also seen how, in *Capital*, he talks about how the planet should be managed as a form of commons.

So when exactly did he turn over a new leaf and break away from productivism? One factor that played a large role in this transition was the work of Justus von Liebig. The criticism of modern agriculture as a "robbery system" that appeared in the seventh edition of Liebig's *Organic Chemistry in Its Applications to Agriculture and Physiology* (more commonly known as *Agricultural Chemistry*), published in 1862, left a deep impression on Marx. He read the book over the course of 1865 and the following year, and it was shortly after that, in 1867, that he worked on publishing the first volume of

Capital. Roughly twenty years had gone by at that point since the publication of the *Communist Manifesto.*

What's key here is that Marx, taking a cue from Liebig, began to develop a theory of metabolism as he worked on *Capital.* Humanity ceaselessly acts upon nature as it produces, consumes, and discards various things as part of conducting life on Earth. Marx refers to this reciprocal cycle as "the metabolic interaction between himself and nature."

Nature, of course, consists of many cyclical processes that are completely independent of humanity. These include photosynthesis, the food chain, and the replenishment of nutrients in the soil. Salmon, for example, swim upstream to spawn. After laying eggs, the salmon die, their corpses breaking down and the nutrients they brought from the ocean ending up in the water and the soil around the streams. Some salmon are eaten by animals like bears and foxes before they have a chance to lay eggs, but these salmon end up becoming nutrients for the soil in the forests via those animals' waste. The leaves that fall from the trees in these forests enrich the soil, and some of them fall into streams to become food for the aquatic insects, crayfish, and other tiny organisms living in the water or to create shade, allowing small fish to hide and grow. The salmon thus enable natural metabolic cycles. Marx refers to these natural cyclical processes as the "metabolism of nature."

Humans, as part of nature, also have a metabolic relationship to the physical world. Breathing is an example of this, as is eating and excreting. Humanity can only live on this planet by participating in these constant cyclical processes of ingestion and excretion as they interact with nature. This is a biologically determined condition for human existence that has been consistent throughout all of human history.

THE METABOLIC DISRUPTION CAUSED BY CAPITALISM

But this isn't the whole story. According to Marx, there's something that ties humans to nature in a way that's distinct from other animals. This something is labor. Labor is an activity unique to humans, mediating and determining "the metabolic interaction between himself and nature."[78]

The point here is that modes of labor differ greatly over time. These differences have major consequences for "the metabolic interaction between nature and man" at different points in history.

Which is to say that capitalism has modified this metabolism in a very specific way. This is due to capitalists striving always to increase their profits. It therefore manipulates the "metabolic interaction between nature and man" so as to always prioritize increasing profit.

Capital is exhaustive in its use of both humans and nature. It mercilessly drives people to work long hours while extracting the power and resources of nature worldwide. Of course, it also spurs technological innovation, developing and introducing new ways to maximize the efficiency of humanity and nature. This increased efficiency has created hitherto unheard-of levels of wealth.

After a certain point, though, it's the negative effects that come to the fore. Capitalists always seek to create as much value as they can as quickly as possible. This leads to capital disrupting the metabolic link between humanity and nature. Physical and mental ailments stemming from excessively long hours of arduous labor are manifestations of this disruption, as are ecological destruction and exhaustion of natural resources.

The metabolism of nature is an ecological process that, at its core, exists independently of capital. Capital, though, has steadily changed this process to suit its needs. As we've seen, capital's unlimited movement to increase its own profits has proven itself to be constitutionally incompatible with the cycles of nature. This incompatibility is the most fundamental cause of the present climate crisis, the Anthropocene itself being its ultimate consequence.

THE IRREPARABLE RIFT

Marx warns us in *Capital* that capitalism opens up an "irreparable rift in the interdependent process of social metabolism." This is in the section touching on Liebig's work as he analyzes how large landed property ownership underpins the business of agriculture under capitalism:

> [In] this way [large-scale landownership] produces conditions that provoke an irreparable rift in the interdependent process between social metabolism and natural metabolism prescribed by the natural laws of the soil. The result of this is a squandering of the vitality of the soil, and trade carries this devastation far beyond the bounds of a single country.[79]

Marx is sounding the alarm here, in the third volume of *Capital*, that capitalism will undermine the conditions necessary for sustainable production by disrupting and opening up a rift in the metabolism connecting humans and nature. Capitalism makes the sustainable management of this metabolic connection impossible and thus hinders the further development of society.

We see here, as well as in *Capital* overall, a complete absence of any sort of uncritical celebration of the productive force of modernization. Rather, we see a clear repudiation of the development of technology and productivity to further the unlimited profit seeking of capital as nothing more than "a progress in the art, not only of robbing the worker, but of robbing the soil."[80]

MARX'S DEEPER RESEARCH INTO ECOLOGY AFTER *CAPITAL*

The fact that Marx spoke of a metabolic rift opened up by capitalism is something any decent introduction to *Capital* and Marxist concepts will mention these days. But the ecological thinking of the later Marx does not stop with the citation of Liebig's critique of modern agriculture as a "robbery system." In the approximately five-year span between the publication of the first volume of *Capital* and his death in 1883, Marx published almost nothing but was engaged the entire time in an intense study of the natural sciences.

As mentioned already, by collecting previously unpublished manuscripts and research notes into a new set of collected works by Marx and Engels, the MEGA project has unearthed Marx's hitherto buried ecological critiques of capitalism.

The range of Marx's late-life study of the natural sciences is astounding. I've analyzed this newly uncovered material in detail in my book *Karl Marx's Ecosocialism*, showing how his notes reveal that his research led him beyond Liebig's critique of modern agriculture as robbery. He addressed ecological issues as diverse as the overharvesting of forestland, the waste of fossil fuels, and the extinction of seeds as arising from the contradictions of capitalism.[81]

A CLEAN BREAK FROM PRODUCTIVISM

The scholar Marx studied most extensively during the course of his research into ecology after the publication of volume one of *Capital* was the German agriculturist Karl Fraas.

Fraas's book, *Climate and the Vegetable World throughout the Ages, a History of Both*, describes the process by which various ancient civilizations fell, including Mesopotamia, Egypt, and Greece, among others. According to Fraas, a common reason all these civilizations collapsed was a change in climate caused by overharvesting forestland, which made the cultivation of the land more and more difficult. Those regions are arid these days, but this was not always the case. Large amounts of fertile land were lost to the overdevelopment of the natural world.

Fraas warned that the rise in temperature and drying of the air associated with overharvesting forests had a huge effect on crop yields, enough that it could lead to civilizational collapse. He was also worried about the danger of capitalism's drive to develop logging and transportation technologies, which would only further extend humanity's incursion into the world's forests.

Marx praised Fraas's book, even stating that it displayed a "socialist tendency."[82] He saw this tendency in Fraas's criticisms of the plunder of nature under capitalism and his call for a sustainable relationship to the world's forestland. Marx made these observations in 1868, the year after the publication of *Capital*'s first volume.

Marx was also aware of the work of William Stanley Jevons, originator of the Jevons Paradox we examined in chapter 2. Jevons warned, touching on the work of Liebig as well, that the easily accessible deposits of coal in England at the time were shrinking. Marx's study of the natural sciences, including geology, led to him paying much attention to the problem of human activity leading to seed extinction as well.

This research was Marx's way of investigating the metabolic rifts opening up in various regions all over the world. It was his intent to examine these rifts and show that they demonstrated, by their very existence, the central contradiction within capitalism.

Marx's stance, as revealed by looking at these late-period notes, is completely different from an optimistic view that sees an increase in productivity as the key to conquering nature and eventually overcoming capitalism. It's obvious at this juncture that he made a clean break with productivism. However, he had not converted completely to a simplistic argument that environmental crisis leads directly to civilizational collapse, either.

Rather, what Marx focused on most closely in the years following the publication of *Capital*'s first volume was the relationship between capitalism and the natural environment. Capitalism buys time by externalizing the metabolic rifts it opens up through technological innovation. All this process of displacement ends up accomplishing, though, is widening the scope of this "irreparable rift" to global proportions. When all is said and done, not even capitalism itself will be able to survive it.

As we saw in chapter 1, Marx spent his later years mapping out in concrete terms exactly how the endless cat-and-mouse game of displacement played out at different points in history.

TOWARD AN ECOSOCIALISM THAT FOSTERS SUSTAINABLE ECONOMIC DEVELOPMENT

In the period following the publication of *Capital*'s first volume, Marx broke away from his earlier superficial celebration of productivism. He then proceeded to examine various sources, looking for a viable path to a form of socialism that could foster economic development in a sustainable manner. Marx was, at this point, convinced that sustainable growth was impossible under capitalism, that capitalism would lead to nothing but the further intensification of nature's plunder. In other words, blindly accelerating productivity under capitalism would never pave the way for a transition to socialism. This is one major shift in Marx's thinking during this period.

Rather than calling for raising productivity under capitalism, Marx now sought first to bring about a transition to a separate economic system—that is, socialism—and then foster sustainable economic development within that system. The vision he pursued during this period was one of ecosocialism.

By the end of his life, though, Marx had progressed yet further, moving beyond even ecosocialism in his analysis.

SHAKING THE FOUNDATIONS OF THE PROGRESSIVE VIEW OF HISTORY

The shift to espousing a viewpoint calling for sustainable development under ecosocialism was of course a major one. But Marx's break from productivism entailed questioning an even broader and more pervasive worldview: the view of history as progress. This will have major consequences for my ultimate argument about Marx's relevance to the Anthropocene.

First, let's review. According to a Marxist interpretation of the progressive view of history, the primary force driving human history forward is the development of productivity. Therefore, every nation is required to raise its productivity by following the path laid out by Western Europe: capitalistic industrialization. This view of history as progress is based on productivism insofar as it sees increasing productivity as the driving force of history itself. Moreover, this productivism becomes the justification for Eurocentrism.

However, once productivism is rejected, a high productivity rate no longer serves as proof that a society has reached a more advanced stage. Progress in terms of technological development that ends up only being destructive loses its meaning as progress. This is why, after Marx made his clean break with productivism, he found himself forced to reevaluate the Eurocentrism that forms the other side of that coin. Once he rejected both productivism and Eurocentrism, Marx had no choice but to question the progressive view of history itself. This amounts to nothing less than a complete rethinking of historical materialism.

Next, I will trace the process of Marx's questioning and subsequent rejection of the progressive view of history. The first step is to examine his relation to Eurocentrism.

CAPITAL AND EUROCENTRISM

It's actually fairly difficult to tell whether Marx eventually rejected Eurocentrism just by looking at what's been published.

It's true that before the publication of Capital, in the latter half of the 1850s, Marx had already espoused anticolonial positions.[83] Marx publicly took the side of the oppressed with regard to the anticolonial struggle in India, the January Uprising in Poland, and the American Civil War. But this doesn't necessarily indicate that he had rid himself entirely of his Eurocentrism.

What about in Capital? In the first edition of the first volume, where we've already detected traces of an ecological point of view, we also find this:

The country that is more developed industrially only shows, to the less developed, the image of its own future.[84]

This sort of simple view of history as progress is undeniably Eurocentric. It's nothing more than the wilful projection of European history onto the rest of the world.

If this were the whole story, this would mean that Marxism could even, in the worst-case scenario, be used to justify colonialism as a way to bring civilization and modernization to "primitive" peoples. It is for this dangerous Eurocentrism that Marx's thought has been repeatedly criticized.

THE ORIENTALISM OF EARLY MARX— SAID'S CRITIQUE

The most famous critique in this regard comes from the world's foremost postcolonial scholar, Palestinian American professor Edward Said. Said calls Marx an Orientalist—a European who sees non-Europeans as lesser and primitive:

> In article after article, he returned with increasing conviction to the idea that even in destroying Asia, Britain was making possible there a real social revolution . . . Marx's humanity, his sympathy for the misery of people, are clearly engaged. Yet in the end it is a Romantic Orientalist vision that wins out.

Said thus concludes that ultimately, "Marx's economic analyses are perfectly fitted . . . to a *standard Orientalist undertaking.*"[85]

What Said is criticizing here is a notorious article Marx wrote for the *New York Tribune* while he was still in his thirties on "The British Rule in India." In the piece, Marx writes:

> England, it is true, in causing a social revolution in Hindostan, was actuated only by the vilest interests, and was stupid in her manner of enforcing them. But that is not the question. The question is, can mankind fulfill its destiny without a fundamental revolution in the social state of Asia? If not, whatever may have been the crimes of England she was

the unconscious tool of history in bringing about that revolution.[86]

Marx of course acknowledges the cruel nature of British colonial rule in India. But he also appears to justify it by evoking its eventual effect on the progress of human history.

At the time, Marx saw Asian society, with India as the prime example, as stagnant, passive, and possessing "no history at all."[87] It therefore took the incursion of a capitalist nation like Britain to drive history from the outside. As Said points out, this displays all the characteristics of an Orientalist point of view. Marx is justifying the suffering produced by history's advancement, calling it a necessary evil for the progress of human history as a whole. As an Asian Marxist, I must say that such a Eurocentric view is utterly unacceptable.

In a draft manuscript for *Capital* written in the 1860s, as part of his critique of the Swiss socialist economist Jean Charles Léonard Simonde de Sismondi, Marx wrote:

> To oppose the welfare of the individual to this end, as Sismondi does, is to assert that the development of the species must be arrested in order to safeguard the welfare of the individual, so that, for instance, no war may be waged in which at all events some individuals perish. Sismondi is only right as against the economists who conceal or deny this contradiction. Apart from the barrenness of such edifying reflections, they reveal a failure to understand the fact that, although at first the development of the capacities of the human species takes place at the cost of the majority of human individuals and whole human classes, in the end it breaks through this contradiction and coincides with the development of the individual; the higher development of individuality is thus only achieved by a historical process during which individuals are sacrificed.[88]

Promote production, even at the expense of the individual! Spread market capitalism throughout the world! Only this can bring about freedom and liberation! It's as if in these passages, Marx has become a neoliberal ideologue.

LOOKING TO NON-WESTERN AND PRECAPITALIST SOCIETIES

Said's criticism of Marx, however, fails to include his later thinking, and so in that sense, it is limited. The new materials brought to light by the MEGA project clarify Marx's profound rethinking of his earlier Eurocentrism. This decisive turn in his thought process began in 1868, right after the publication of volume one of *Capital*.

The year 1868 marked the beginning not only of Marx's serious research into the natural sciences and ecology but also into the land use and agricultural practices of non-Western and precapitalist societies. He read extensively on ancient Rome, Indigenous peoples in America, India, Algeria, and South America. He also took great interest in agricultural communes in Russia, even teaching himself Russian in order to examine materials on cooperative land management and cultivation.

Looking at his research notes from this time, we see Marx not only criticizing Britain for its colonialism but also writing positively about the tenacious resistance displayed by Indian farming communes. This is clearly a different Marx from the one who wrote "The British Rule in India" in 1853.

MARX'S CLEAN BREAK FROM EUROCENTRISM—THE LETTER TO VERA ZASULICH

Marx expressed this change in his thinking most clearly toward the end of his life, when he participated in a debate about the future of Russian communes. In 1881, two years before he died, Marx wrote a letter to the Russian revolutionary Vera Zasulich.

Among the materials Marx wrote in the fourteen years following the publication of the first volume of *Capital*, this letter shows us most clearly Marx's explicit critique of the progressive view of history, demonstrating just how much his viewpoint had changed. Indeed, it isn't too much to say that this letter holds within it the ultimate culmination of Marx's ideology.

At the time, Russia still had agricultural communes called *mirs*. Some Russians, called the Narodniks, argued for a socialist revolution that would topple imperial domination by expanding these communes. There was an intense debate at the time among Russian revolutionaries about whether Russia could attain socialism without first passing through the capitalist stage of development.

The problem here was the line from *Capital*'s first volume, quoted earlier. Let's look at it again.

The country that is more developed industrially only shows, to the less developed, the image of its own future.

The crux of the debate was whether this quote applied to Russia— that is, did Russia, as this sentence implies, have to modernize itself under capitalism before achieving a socialist revolution? Zasulich decided to ask Marx what he thought.

Marx's response to her letter is, in fact, rather curt. But he wrote and rewrote at least three much longer drafts of the letter before sending it. This was obviously due to Zasulich's inquiry cutting right to the heart of the problem for Marx. After all, this was a non-Western person asking, four years after the publication of *Capital*'s first volume, whether the Eurocentric view of history as progress that it puts forth was actually correct.

In his reply, Marx clearly states that the historical analysis in *Capital* is confined to Western Europe. There is no need to destroy the agricultural communes left in Russia to spur modernization. In fact, in the case of Russia, these communes play an important role in resisting the force of capitalism as it seeks to swallow up the whole world. There is a chance, Marx writes, for Russia to attain communism by developing these communes "in their current state" while reaping the positive effects of capitalism as it evolved in the West.

The important thing here is Marx's clear articulation that Russia could transition to communism "without passing through the Caudine Forks" of a capitalist stage of development.[89] It's clear here that late in his life, Marx made a clean break from Eurocentrism and the simplistic view of history that comes with it.

THE EVIDENCE OF THE "PREFACE TO THE SECOND RUSSIAN EDITION OF THE *COMMUNIST MANIFESTO*"

The same idea can be found in the "Preface to the Second Russian Edition of the *Communist Manifesto*," which was published the following year. In it, Marx and Engels state:

> If the Russian Revolution becomes the signal for a proletarian revolution in the West, so that both complement each other, the present Russian common ownership of land may serve as the starting point for a communist development.[90]

Here Marx is making sure to include praise for communal land-ownership as exemplified by the *mir*. This isn't just lip service for his Russian readers. Without this new preface, the *Manifesto* he cowrote in his youth, with its celebration of history as progress, would be open to terrible misinterpretation. Marx, now near the end of his life, understood this danger well and thus made sure to conclude the new preface with this sentiment.

Furthermore, this preface clearly asserts that not only do Russian communes not have to pass through a capitalist stage in development but that the country may actually be ahead of Western Europe—even if a complementary revolution in the West is still necessary—and that the transition to communism can be affected there first. It's undeniable, looking at this, that Marx's view of history has undergone a fundamental transformation.

There is no reason to confine this argument to Russia. The expansion of communes in Asia and Latin America should also be able to play the same role.

And in fact, Marx did see not only the Russian *mir* but the "village communes" of Asia as types of archaic communal forms that had survived to the present while avoiding violent destruction by capitalism. In other words, communes all over the world held the same potential as the Russian ones.

These writings are why University of California sociologist Kevin Anderson asserts that the late Marx embraced a multilinear historical view, refusing to "bind himself to a single model of development or revolution" dependent on a unilinear view of history as progress.[91]

The road to socialism is thus unbound from the developmental model of Western Europe. Rather, Marx concludes that the search for methods for transitioning to a communist society must include the thorough study of the complexities and differences in the histories and social systems of non-Western societies.

Marx's Eurocentric view of history as progress thus ends up replaced by one that focuses on non-Western societies, actively praising the communal forms found in them. And if this is true, then Said's criticism of Marx as an Orientalist is no longer sustainable.

THE NEW FACE OF MARX'S COMMUNISM

Did Marx's change in thinking late in his life stop with the abandonment of his former view of history as unilinear progress? As a matter of fact, it didn't.

Even Professor Anderson, who sings the praises of late Marx's study of worldwide communes in his book *Marx at the Margins*, fails to grasp the true meaning behind this shift. It's my contention that the theoretical importance of Marx's letter to Zasulich exceeds what is indicated by Anderson.

Pointing out that the late Marx rejects the progressive view of history is itself no longer a new revelation, as experts have been doing this for more than a decade now.[92] As previously mentioned, Marx had also made his anticolonial views clear in the latter half of the 1850s and from that point on continually stressed the importance of anticolonialism to anticapitalist struggle.

If Marx's late-period shift in thought over almost twenty years' time, including such intense research into world communal forms, consisted merely of the abandonment of Eurocentrism and the adoption of a multilinear view of history, that would be rather unsatisfying, wouldn't it?

Here I want to push things further. The question of whether Marx abandoned a unilinear progressive view of history is only the first step toward readying my ultimate reading of Marx's utility to the present moment. The real question here is: What insights came to Marx after he changed his views on history and progress?

To answer this question, though, it's insufficient to simply point out that Marx's study of communal forms led to his abandonment of his former Eurocentrism. If the ultimate conclusion of Anderson's otherwise brilliant research feels a bit lackluster, it's because it stops at merely highlighting Marx's abandonment of one facet— Eurocentrism—of his view of history. We must combine this insight with analysis of the other part of Marx's view of history to clarify its importance: his shift away from productivism. It's only by thinking about his break from Eurocentrism in conjunction with the break from productivism occasioned by his research into ecology that the true nature of Marx's shift in thought comes into focus.

When combined with the problem of productivism, an even more surprising interpretation arises than the simple shift from a singular path to multiple possible paths to reach the promised land of communism. In short, it represents a fundamental change in the nature of Marx's conception of communism itself, its new face revealed at the end of his life. This is the aspect of Marx's thinking that has not been fully investigated previously. We are reaching the main idea at last.

WHAT DELAYED MARX'S COMPLETION OF *CAPITAL*?

The extraordinary delay of his completion of the second and third volumes of *Capital* can be explained by the possibility of a fundamental change in Marx's communism toward the end of his life. Engels pushed really hard for Marx to complete *Capital*, but sixteen years after the publication of the first volume, Marx passed away without having done so. As we've seen, Marx spent those years studying ecology and communal forms. Why was Marx preoccupied with this research to the point that he couldn't progress in his writing? On the surface, it might seem that Marx, plagued with various illnesses, immersed himself in tangential reading as an escape from the hard work of completing *Capital*.

But this is not the case. Look at how metabolism became a new theoretical axis of Marx's thought. On top of this, we see Marx sweating blood as he moves away from his convictions about history as progress, arriving finally at a completely new conception of history. His research into ecology and the communal societies of the non-Western and precapitalist world was absolutely essential to constructing his new vision of revolution.

They may seem unrelated at first glance, but together, these two lines of research form the fundamental framework underlying the ultimate destination of his thinking. So what is the relation between the two?

CIVILIZATIONAL COLLAPSE AND COMMUNAL SURVIVAL

Let's start by examining why Marx was so preoccupied with research into communes at the end of his life. The spark of this interest came from his encounter at the beginning of 1868 with the previously mentioned book by Fraas during his research into ecology. In this sense, his research into communal forms was connected to his research into ecology from the very start.

We previously looked at Marx's reading of Fraas's views on civilizational collapse, but it's important to recognize that Fraas also touched on communal societies that didn't follow these civilizations down the road to destruction.

Fraas praises especially the *Markgenossenschaft* ("Mark-associations") of ancient Germanic peoples for their sustainable farming practices. While these Germanic tribes are usually thought of as "barbarians," from the point of view of sustainability, they seem to have been quite advanced. *Markgenossenschaft* is a broad term for the Germanic tribal societies Tacitus wrote about during the reign of the Caesars in Rome. This period saw a shift from tribal communities focused on hunting and military matters to sedentary agricultural communities.

These Germanic peoples owned land communally and had strong rules regulating production methods. It was unthinkable within the *Markgenossenschaft* to sell land to anyone outside the community. Other products like timber, pork, wine, and the like were also forbidden to be bought and sold outside the community.[93]

This sort of strong communal regulation supported the renewal of the soil and enabled sustainable farming practices. It even brought about long-term improvements in soil fertility. It was a very different arrangement from those of ancient civilizations that had weakened these sorts of communal bonds. And of course, it's the complete opposite of agribusiness under capitalism, which includes the extraction of nutrients from the soil in order to make money selling crops in large urban centers.

Marx immersed himself in all of Fraas's writings. The ecological viewpoint evident in *Capital* led to a growing interest in the sustainability of communal forms in precapitalist societies.

ENCOUNTERING EGALITARIANISM IN THE COMMUNE

Marx's strong interest in Fraas's discussion of *Markgenossenschaft* can also be seen in his careful reading of papers on these communities written by German legal historian Georg Ludwig von

Maurer. It was Maurer's work that formed the underpinning for Fraas's work on *Markgenossenschaft*. Tellingly, Marx detected "socialist tendencies" in Maurer, just as he did in Fraas.[94]

This is due to the following features of Maurer's argument. According to him, *Markgenossenschaft* involved not only communal landownership that allowed everyone to cultivate the land equally. Members of the commune also regularly exchanged plots of land to work, each allotment being decided by lottery. Maurer draws attention to how this allowed them to avoid a situation in which the most fertile land belonged to only part of the population, thus preventing unequal wealth distribution.

This arrangement provides a stark contrast to the *landifundia* of ancient Rome, vast estates presided over by nobles who managed them using slave labor. What Maurer, a conservative thinker, unearthed in his historical study was a form of egalitarianism practiced by the very same Germanic "barbarians" who struck fear in the hearts of even the socialists of Marx's time.[95]

THE BASIS OF A NEW COMMUNISM— SUSTAINABILITY AND SOCIAL EQUALITY

Of course, Marx recognized the egalitarianism of communal societies even before 1868. Marx even used the phrase "indigenous communism" to describe these archaic communal forms in *Capital*.[96]

But behind Marx's use of the same phrase—"socialist tendencies"—to praise both Fraas and Maurer right after Marx published *Capital*'s first volume, lies a completely new realization on his part, one he'd begun contemplating seriously for the first time: that sustainability and social equality were intimately linked. This is the real reason why Marx began to study non-Western and precapitalist communal forms alongside his research into ecology.

The Germanic peoples treated land as shared property. It belonged to no one. They prevented monopolization of the blessings of this land by the few by dividing the land equally among everyone. By avoiding the individual hoarding of wealth, domination and subservience were prevented from arising among the members of the commune.

At the same time, because the land belonged to no one, it was protected from overuse by a single owner. This guaranteed that the land was used sustainably.

This is the intimate connection between social equality and sustainability. It is precisely this intimate connection that inoculated these communes against capitalism and seemed to pave the way for them to transition directly into communism. Marx would become more and more interested in exploring this possibility.

REREADING THE LETTER TO ZASULICH— AN ECOLOGICAL VIEW

The culmination of this line of thought can be found in Marx's letter to Vera Zasulich. Let us examine the text of this letter in more detail.

First, Maurer and his research into communal forms make an appearance in the letter. Marx states that the archaic communes still extant in Russia are of the same nature as the "agricultural communes" found in Western Europe in the form of the communal arrangements of these Germanic peoples.

Marx continues by saying that the "natural viability" of these agricultural communes is very strong. While other sorts of communes disappeared with the vicissitudes of constant war and migration, agricultural communes survived until the Middle Ages. Marx points to the survival of shared ownership of forests and farmland in the Trier district in his home country of Germany as a trace of German tenacity.

Indeed, Marx praises this form of social arrangement and its survival into the Middle Ages by writing to Zasulich that:

> The new commune—in which cultivable land is privately owned by the producers, while the forests, pastures, waste ground, etc., still remain communal property—was introduced

by the Germans to all the countries they conquered. Thanks to certain features borrowed from its prototype, it became the only focus of popular life and liberty throughout the Middle Ages.[97]

It was based on this praise for these communes that Marx assured Zasulich that he didn't want to force Russia to take the road to modernization via capitalism.[98] If agricultural communes were left in their original form in Russia, the transition to communism could take place there based on their power. This argument demonstrates the large shift that had occurred in Marx's view of history as well at that point.

What's even more important to note here, though, is his consciousness of the problem of ecology. We can see in this letter the following features of Marx's later thought. He recognized that raising productivity under capitalism doesn't necessarily lead to the liberation of humanity. Indeed, it disrupts and eventually creates a rift in the metabolic link between humans and nature that forms the base conditions for life itself. Capitalism does not bring about progress toward communism. Rather, capitalism destroys the "natural viability" necessary for society to thrive. This is where Marx's thinking stood at this point.

This newly conceived argument demanded that Marx reassess his earlier views on history as progress. If capitalism brings about not progress but the irreversible destruction of the natural environment and the devastation of society, this shakes the very foundations of linear historical thinking. It is no longer self-evident that Western Europe, with its high level of productivity, is superior to the non-West.

As we've just seen, Fraas and Maurer assert that *Markgenossenschaft*, as a form of social organization mediating the metabolic link between humans and nature, fostered more sustainable and equal relationships between people and between people and nature than later forms of organization. In this sense, though its productivity was comparatively quite low, *Markgenossenschaft* was in fact "superior."

Such a major revision of Marx's theoretical framework naturally made completing the second and third volumes of *Capital* extremely difficult. Yet such a fundamental rethinking of his conceptualization of history had to be incorporated into any new writing he did to complete the *Capital* project. This rethinking also necessitated further research into non-Western and precapitalist communal forms, as well as into the natural sciences with a focus on ecology.

CAPITALISM'S BATTLE WITH ECOLOGISTS

Marx's repudiation of the progressive view of history led to a major shift in his analysis of the situation in Western Europe, including Britain, where he resided at the time. This seems only natural. After all, he wasn't studying communes on a whim—he was doing so precisely to overcome Western European capitalism.

This facet of his shift in thought can be seen in the letter to Zasulich as well, in a section about the dangers of capitalism in the West:

Today, [capitalism] faces a social system which, both in Western Europe and the United States, is in conflict with science, with the popular masses, and with the very productive forces that it generates—in short, in a crisis.[99]

This section's language about capitalism "in conflict with science" has been, up till now, interpreted by Marxist-Leninists speaking from a productivist standpoint to mean that yet more productive force was necessary to spur development and eventual revolution. In other words, increasing productive force was the way to overcome the crisis caused by capitalism.

It's for this reason that Marx's famous definition of communism from the *Critique of the Gotha Program*, "from each according to his abilities, to each according to his needs!"[100] was interpreted as a call for unlimited productivity to produce "unlimited abundance" and thus solve the problem of unequal distribution.[101]

But looking at that line from the Zasulich letter as a criticism of productivism that builds upon his analysis of the metabolic rift opened up by capitalism, we can see that it means the exact opposite. Marx argued for neither limitless growth nor infinite mass production.

The "science" with which Western European societies are in "conflict" is none other than the science focused on the world's environment as practiced by Liebig and Fraas—namely, ecology.

Ecologists, as they developed their critiques of capitalism's plunder, were shaking the foundations of any justification for capitalism. Their science exposes the failure of the productivist project to conquer nature through technology and thus free humanity from its limitations.

What Liebig and thinkers like him made abundantly clear was that capitalism could not raise productivity any higher without sacrificing sustainability. To heedlessly push productivity higher anyway amounted to the plunder of the environment. And that's not all—it would also destroy nature's ability to renew itself. This sort of capitalism is impossible to justify or continue.

Now that we've seen Marx's movement toward ecological thought, there's no other way to interpret this line about capitalism "in conflict with science."

THE NEW RATIONALIZATION—TOWARD THE SUSTAINABLE MANAGEMENT OF THE EARTH

What Marx took from Liebig and Fraas was a new viewpoint, one that argues for a "rational" agriculture based on the insights of the natural sciences that would overcome the dangers posed by capitalism. The rationalization Marx speaks of here is, of course, very different from efficiency under capitalism, which seeks only to maximize profits. It's a "new rationalization."

In part six of the third volume of *Capital* edited by Engels after Marx's death, which deals with "ground rent," Marx talks about the *irrationality* of land use under capitalism:

[I]nstead of a conscious and rational treatment of the land as permanent communal property, as the inalienable condition for the existence and reproduction of the chain of human generations, we have the exploitation and the squandering of the powers of the earth.[102]

Capitalism uses the natural sciences only to extract nature's force without compensation. As a result, rising productivity only intensifies plunder, undercutting the foundation necessary for human development to maintain its sustainability. From a long-term point of view, such use of the natural sciences can only result in "exploitation" and "squander" and can never be "rational."

Marx, as he launches this critique, is calling for the sustainable management of the land—that is, the Earth—as a form of commons. It's precisely this that makes up the more "rational" economic system called for by Liebig and Fraas as well.

This sort of demand founded on science exposes the essential irrationality of capitalism, bringing its legitimacy into "crisis."

In his letter to Zasulich, just after the line quoted earlier, Marx concludes:

> The best proof that this development of the "rural commune" is in keeping with the historical trend of our age is the fatal crisis which capitalist production has undergone in the European and American countries where it has reached its highest peak, a crisis that will end only when the social system is eliminated through the return of modern societies to the "archaic" type of communal property.[103]

This is not an assertion that capitalism advancing to its maximum level of development will result in the emergence of communism. Rather, it's that the *Markgenossenschaft* of the Germanic peoples and the *mirs* in Russia hold within them elements to which the modern societies in Western Europe must return.

So what exactly are these things that the Global North must learn from the *mir* and the *Markgenossenschaft* so as to revive them in the present age?

THE REAL THEORETICAL SHIFT—A TRANSFORMED COMMUNISM

Now, at last, we have reached the heart of the matter. I would like to summarize the argument I've made so far and then present my ultimate conclusion.

First, what allowed Marx, starting in 1868, to discard the progressive view of history he'd previously embraced was his research into the natural sciences and communal forms of social organization. We can see the culmination of these two strands of research in his firm belief that they were, in fact, intimately linked, a theoretical achievement we can first observe in what stands as the high point in Marx's thinking late in life, the letter to Zasulich.

In other words, Marx's research into the natural sciences and communal forms of social organization led him to deepen his reflections on the links between sustainability and equality. As he wrote and rewrote the letter to Zasulich, he was attempting to develop a new form for rationalization, one that would stand as the goal for any future society. In short, the question posed by this Russian activist provided the opportunity for Marx to reimagine the ideal way to bring about a form of Western European society founded on principles of sustainability and equality.

What finally emerges from this intellectual project is the real shift in Marx's thought at the end of his life. The break he makes from the progressive view of history, which began with his research into ecology, leads to a fundamental rethinking of his assumptions about the "advanced" nature of Western European capitalism.

This leads not only to acknowledging that there are multiple roads that might lead to communism but to the reimagination of—and therefore, a major transformation in—the communism toward which Western European capitalism must now aspire.

Let us now look at what this might be.

Traditional communes are based on completely different principles of production from capitalism. The communes described by Maurer and Fraas were defined by strong internal social regulation, completely separate from the logic of commodity production found in capitalism. For example, recall that not only land but even goods produced from the land could not be sold outside the *Markgenossenschaft*.

These communes all shared the same types of traditions related to production. In short, they had cyclical, steady-state economies that weren't designed to grow.

The communes didn't suffer from low levels of production and poverty stemming from their "underdevelopment" and "ignorance." Rather, at moments when they could have worked harder and longer and raised their levels of production, they simply chose not to. And they thus avoided creating the kinds of power dynamics that would evolve into domination and subservience.

MARX'S MOVEMENT TOWARD DEGROWTH

What's important to note here is Marx's affirmation that the stability of a communal society detached from economic growth would foster a metabolic relationship between humans and nature that would be both sustainable and equal.

As we've seen above, Marx asserted at the beginning of the 1850s that communal societies in India were passive, stagnant, and "completely devoid of history" precisely due to their steady-state economies. Such a statement reads like a condensation of productivism and Eurocentrism.

However, Marx ends up asserting that it's precisely the steady state of a commune's economy that allows it not only to resist colonial domination but also to hold within it the possibility of toppling the power of capital and achieving communism. There is clearly a huge shift here. The commune is active in its resistance and holds the power to shape history—the power to bring about communism. This is an affirmation of steady-state economics that stands in total opposition to Marx's writings in the 1850s.

It is Marx's late-period study of ecology that enabled him to see the potential in communal societies. In other words, Marx's interest in sustainability is what brought him around to seeing communes in very different ways than he had in the 1850s. Marx's ultimate goal was not simply finding a road to historical development for non-Western societies like Russia. Indeed, we might even see the development of a multilinear path for historical development as a mere by-product of the true shift in his thinking. Marx's focus was always on constructing a future Western European society. This is why he applied himself to the study of communal forms of social organization.

It was after more than fourteen years of this study that Marx concluded that the sustainability and equality characteristic of a steady-state economy had the power to resist capital and provide the foundation for the society of the future.

It's sustainability and equality that Western European societies need to consciously revive so as to overcome the crisis of capitalism; the material condition necessary to bring this about is a steady-state economy. In short, the communism envisioned by Marx late in his life was an egalitarian, sustainable form of degrowth economics.

When Marx states that Western Europe must "return . . . modern society to a higher form of the most archaic type" to overcome the dangers of capitalism, he was calling for the reinstatement there, at a high level, of the communal principle of steady-state economics.

DEGROWTH COMMUNISM AS THE ULTIMATE GOAL

At this point, the meaning of Marx's use of "return" should be clear. He means that in Western Europe, the effort to bring about communism must consist of learning about the principles of steady-state economics as developed in communal societies and reinstating them in order to bring about a new form of rationalization based on sustainability and equality.

One thing we must remember here is that this form of thought is not a nostalgic call to "return to the village!" or "form a commune!"—Marx repeatedly said that the Russian *mirs*, for example, must incorporate the positive aspects of the technological revolution brought about by capitalism. The revolution he sought in Western Europe would preserve the achievements of modern society even as it called for a form of communism of "the most archaic type"—that is, one that uses a steady-state society as its model.

This means that communism like the Soviet Union's, founded as it was on productivism and the promotion of economic growth, will never work. The Marx we see here would even condemn such a thing as the promotion of capitalistic principles in a way that would never lead to the achievement of the society of the future.

To repeat, this position is the exact opposite of the one held by the young Marx. This is also different from the ecosocialism of Marx when he was influenced by Liebig while writing *Capital*. At that time, he still believed that sustainable economic growth could be achieved under socialism. But in the end, he discarded even this idea.

		Growth	Sustainability
1840s–1850s	**Productivism** *Communist Manifesto,* "British Rule in India"	O	X
1860s	**Ecosocialism** *Capital,* Volume One	O	O
1870s–1880s	**Degrowth Communism** *Critique of the Gotha Program,* Letter to Vera Zasulich	X	O

Figure 8. The evolution of Marx's political goals

In this way, Marx's vision of the future of society underwent a truly major shift at the end of his life. To borrow a once-fashionable turn of phrase from Louis Althusser, it's a shift thoroughgoing enough to be called an "epistemological break."

In short, Marx, having discarded a progressive view of history, incorporated the principles of sustainability and steady-state economics from communal societies into his revolutionary thought. As a result, his vision of communism became something different from both productivism and ecosocialism. What Marx achieved at the end of his life was a vision of *degrowth communism*.

This is the shape of society's future that no one had pointed out before, that constitutes the truly new analysis presented here. Even Engels, his comrade, was unable to grasp it. As a result, Marx's ideas were misunderstood after his death as founded on a unilinear view of history as progress—that is, as a leftist paradigm for productivism.

This is why during the 150 years that have passed since the publication of *Capital*'s first volume, Marxism couldn't be used to analyze environmental problems as stemming from the ultimate contradiction within capitalism, and the crisis of the Anthropocene has been allowed to worsen to its present state.

THIS NEW WEAPON CALLED DEGROWTH COMMUNISM

Up until now, Marxism and degrowth have been thought of as irreconcilable, like mixing oil and water. The conventional understanding of Marxism held that communism imagined a future society brought about when the workers seized the means of production and, once free, unleashed the forces of productivity and technology to provide themselves with affluent lifestyles. Such a society would be incompatible with degrowth.

For this reason, even though it was well known that Marx studied communes and ecology later in life, no one thought to put these strands of enquiry together. This was due to Marxists' inability to accept degrowth as a viable option.

Researchers of course welcomed work like Kevin Anderson's that showed how Marx turned away from his early Eurocentrism. One might say that such work allows Marx to edge closer to a contemporary version of political correctness. When I published my book *Karl Marx's Ecosocialism* (Monthly Review, 2017), revealing what one might call an "environmentalist Marx," it was received by the world's Marxists in much the same way—as a further burnishing of Marx's PC bona fides.

No one, though, went so far as to propose degrowth communism. Even in *Karl Marx's Ecosocialism*, I stopped at noting how a call for sustainable economic development was part of Marx's ecosocialist thought.

We can see in this the heavy burden productivism has placed on Marxism's legacy. Marxism has historically been unable to accept that increasing productivity itself is destructive and so has always seen degrowth as an enemy.

But the fact is, Marx's change of mind after turning away from productivism and concentrating on searching for revolutionary possibility in the study of non-Western and precapitalist communal societies is not just evidence that we should revise the way we imagine him. The radicality of Marx's rejection of productivism and Eurocentrism late in life is not simply a way to make Marx PC and thus more palatable to today's audiences.

Rather, it's evidence that Marx reached the point of imagining *degrowth communism* as a project that might truly topple Western European capitalism.

The above analysis is not meant simply to clarify the precise nature of communism sketched out near the end of Marx's life. It's meant to show that the place Marx reached allowed him to see an entirely new idea come into being, one that no one else had ever articulated before, a new weapon with the power to construct a whole new future for society. A weapon called degrowth communism.

A NEW READING OF THE *CRITIQUE OF THE GOTHA PROGRAM*

Just how far-fetched is this argument? I assert that it's not far-fetched at all.

I want to revisit the *Critique of the Gotha Program*, a piece mentioned previously that Marx wrote near the end of his life. This text discusses the nature of revolution in Western Europe. I want to draw attention to a particular term that comes up in it: "cooperative wealth." This turn of phrase appears in a famous section in which Marx argues that when people are liberated from the domination of capital and regain their ability to labor freely, the nature of wealth itself transforms:

> In a higher phase of communist society, after the enslaving sub-ordination of the individual to the division of labor, and therewith also the antithesis between mental and physical labor, has vanished; after labor has become not only a means of life but life's prime want; after the productive forces have also increased with the all-around development of the individual, and all the springs of cooperative wealth flow more abundantly—only then can the narrow horizon of bourgeois right be crossed in its entirety and society inscribe on its banners: From each according to his ability, to each according to his needs![104]

According to Marx, production under communism will transform from something undertaken only to increase an individual's income and assets into something that allows everyone in the future society to jointly manage cooperative wealth (*der genossenschaftliche Reichthum*). To translate this into words more familiar from the beginning of this chapter, it's my opinion that here Marx is talking about the commons.

Marx has used the word "cooperative" (*genossenschaftlich*) from time to time before. The German word *genossenschaftlich* carries with it the connotation of both cooperatives understood as unions and cooperatives understood as free associations, and he tended to use it in phrases like "cooperative production" and "cooperative mode of production."

"Cooperative wealth," though, is a phrase that only appears in this line from the *Critique of the Gotha Program*. If we were to translate it in line with previous instances—as "the wealth of the cooperative," say—it wouldn't really work. The sense of the phrase would become something like "after the productive forces have also increased with the all-around development of the individual, and all the springs of wealth flow more abundantly for the cooperative"—a passage that sounds like nothing other than a further celebration of productivism! Yet this is the Marx of the 1870s. We have no reason to believe that he would maintain such a position at this point.

The fact is, the origin of this particular use of the word *genossenschaftlich* is very likely different from what it had been in his earlier writings. What might it be?

An obvious possible origin for the term as it's used in the *Critique of the Gotha Program* is *Markgenossenschaft*, the term for the communal social organization of ancient Germanic peoples. There's a high probability that Marx's use of the term here came out of his concurrent research into communes, including, as we've seen, *Markgenossenschaft*. If this is the case, then we should understand the term less as "cooperative wealth" than as "communal wealth." It seems quite natural to thus read the sentence as referring to the cooperative management of wealth understood as communal.

Put another way, can't we read this passage overall as stating that the societal cooperation fostered by communism is modeled on the management of communal wealth in the *Markgenossenschaft* and, as such, should be reinstated in Western Europe as well? It should go without saying that the abundance referred to here is of a different order from the abundance created from endless unchecked production. This is a *radical abundance*, one produced via the commons.

This, finally, is the real theoretical shift achieved by Marx at the end of his life.

FULFILLING MARX'S LAST WISH

It is true, however, that Marx left behind no piece of writing laying out exactly what his vision of degrowth communism might be. Nevertheless, as we've seen, it's possible to put together the various resources gathered in the MEGA project relating to Marx's research into communes and the natural sciences and see the heights to which Marx's thought reached by the end of his life.

This is a version of Marx never before contemplated, and overlooking it has led to the stagnation of present-day Marxism and the deepening of the worldwide environmental crisis. Marxism has been bound, from its initial promulgation all the way to the present day, to productivism. Even Marxists who criticize the Soviet Union have not been able to free themselves completely from productivism's shackles.

The dead seriousness of the climate crisis facing modern society, caused as it is by unfettered productivity, means that there is no longer any room left for championing productivism. We saw how difficult decoupling is to put into practice, and with that in mind, even ecosocialism seems insufficient as a way forward.

Capitalism's globalization has expanded to an extent unimaginable in the nineteenth century, its contradictions threatening the continued existence of all of humanity. Now is the time when we must pursue the path of degrowth communism first indicated by Marx. Marx's last wish, written in the letter to Zasulich at the end of his life, is what we must heed if we wish to survive the Anthropocene at all.

The theoretical shift Marx underwent near the end of his life ended up being too big, and he died before he could apply it to the unpublished sections of *Capital*. But the cue we need to create the future society we want is buried there in the territory just beyond the ideas he was able to develop during his lifetime.

For this reason, in order to face the crisis of the Anthropocene, we must develop Marx's critique of capitalism and complete what he started in *Capital* by fully theorizing what degrowth communism might look like, creating a major new analysis adequate to this new age.

5

The Wishful Thinking Known as Accelerationism

TOWARD *CAPITAL* FOR THE ANTHROPOCENE

What has become clear at this point is that the only thing that can adequately address the present age of climate crisis is communism.

As ever-expanding economic activity threatens to destroy the environment completely, we find ourselves at a moment when, if we don't use our own hands to stop capitalism's advance, we will witness the end of human history. It's imperative that we install a social system other than capitalism in this era of climate crisis. Communism is the only viable choice left for a future in the Anthropocene.

The term "communism," though, has various possible definitions. I stand with Marx's definition at the end of his life, gazing out from the apex of his thought to contemplate a form of communism premised on degrowth. But there exists another view, one that calls for bringing about communism via the further acceleration of economic growth. This is what is known as "left-accelerationism," and it has gathered increasing support in recent years.

To put it bluntly, accelerationism is nothing more than an aberration born of ignorance of the insights attained by Marx in his later years. It's a product of the 150-year-long misunderstanding of productivism as Marxism's true essence. It is, however, also a theory whose viability is being seriously debated amongst some who profess to be concerned about the environment.

In this chapter, I intend to investigate and critique accelerationism as something that can teach us, by contrast, about the proper way forward. My hope is that this will aid in making Marx's thinking later in life, as well as the degrowth communism this book proposes, easier to imagine.

WHAT IS ACCELERATIONISM?

Accelerationism calls for the promotion of sustainable growth. It holds that completely sustainable economic growth is possible under communism, which awaits us at the end of the technological advances of capitalism.

Published in 2019, *Fully Automated Luxury Communism: A Manifesto* by young British journalist Aaron Bastani is an example of a book articulating this vision.

Bastani starts with the premise that climate change and population increase are threatening human life at the civilizational level during the twenty-first century. He means that economic development and increasing populations in the Global South place a huge burden on the environment due to the increased volume of their consumption of resources and the increased land area needed for the cultivation of food and other crops. This may all-too-easily lead us into an irreversible state of climate crisis. But even so, people in emerging countries should not simply be made to put up with their lot for the good of everyone else. Bastani points to this contradiction as the crux of the current difficulties facing the global environmentalist movement.

This argument has plenty in common with my own argument in this book. But our opinions on what should be done next could not be more different. According to Bastani, these difficulties can all be resolved using the new technologies now being developed at a prodigious rate.

The technological revolution of today is comparable, Bastani asserts, to historical turning points like the beginning of agriculture or the first use of fossil fuels. It takes a large amount of land to raise a cow, so what should we do? Replace beef with factory-made artificial meat! What about the various illnesses and maladies plaguing humankind? Prevent them by using genetic engineering! Automation will free humankind from the bonds of labor, but how will these marvelous robots be powered? With the free and infinite power of the sun![105]

To be sure, rare metals like cobalt and lithium exist only in limited amounts on Earth. But that's no cause to worry, Bastani assures us. If we just develop the technology necessary to harvest resources from space, we'll be able to take what we need from nearby asteroids. This is a worldview that does not recognize the existence of natural limits.[106]

Of course, new green technologies are not yet capable of being used at that level, and even if they were to be commercialized, they would never be profitable. But Bastani remains optimistic. He puts his faith in Moore's Law, by which technological development accelerates exponentially so that before long, all these technologies will reach the point of concrete application and use.

As usage spreads, Bastani predicts, productivity will rise within the relevant sectors until a revolutionary transformation occurs within the market's price mechanism.[107] This is because a price mechanism only functions when there's scarcity. For example, air exists in abundance, which means it cannot be priced. Sunlight and geothermal energy exist in abundance, unlike fossil fuels, so as soon as the infrastructural costs of accessing these energy sources go down, the energy they produce will become free.

The idea is that if production increases exponentially, the prices of things will continue to drop until we reach a "luxury economy" unbound by natural restrictions, not even those of monetary currency. This is the "fully automated luxury communism" Bastani proposes. It's a future in which no one has to worry about environmental issues any longer and can instead make use of their fortunes freely and without consequences.

This is, according to Bastani, the ultimate expression of "from each according to his ability, to each according to his needs"—the ultimate form of the communism envisioned by Marx.

ECOMODERNISM STRIKES BACK

Bastani's brand of optimistic projection is for me a classic form of the productivism Marx broke away from at the end of his life. This line of thought is now referred to as "ecomodernism." Ecomodernism holds that the unconditional application of technologies like nuclear energy, geoengineering, and negative emissions technologies (the NETs discussed in chapter 2) will allow the Earth and its environment to be "used and operated" in an optimal way. Rather than an understanding of nature's limits leading to coexistence with nature, humanity's survival will be enabled by nature's *management*. The Breakthrough Institute mentioned earlier is a big proponent of this version of ecomodernism.

The problem with ecomodernism is its aggressiveness. It holds that the environmental crisis has become so bad that there's no way to go back. Therefore, we have no choice but to encroach upon nature even more, to manage it even more invasively so as to protect our way of life. To take one example, the French philosopher Bruno Latour has expressed this as a call to "love your monsters," putting forth the ecomodernist idea that we must love our technologies like we love our children and must not reject them as "monsters."[108]

Ecomodernists like Bastani and Latour are engaging in a form of what Johan Rockström calls "wishful thinking." We examined the phoniness of the "green growth" contingent earlier, but the difficulty of decoupling doesn't disappear with the advent of communism—environmental sustainability and unlimited economic growth are two things that can never go together.

If the scope of the economy were to expand by a factor of two or three under a version of Bastani-style accelerationist communism, resource extraction would still necessarily increase. As a result, even if solar energy were to replace fossil fuels completely, the amount of emissions saved would be moot and the volume of total atmospheric carbon dioxide would still go up. Communism is not exempt from the Jevons Paradox.

Left-accelerationism aims to solve world poverty through increased productivity and replacing fossil fuels with other energy sources. Yet, ironically enough, the result of all this would be the increased plunder of the Earth and a potentially even more egregious form of ecological imperialism.

WHOSE POLITICS ARE "FOLK POLITICS"?

This is not the only problem with accelerationism. Not only is it impossible from a scientific point of view, the process it proposes for revolution is full of problems as well.

Accelerationism repeatedly takes the post–Cold War Left to task for the environmental movements associated with it: organic farming, slow food, local production for local consumption, vegetarianism, and so on. They criticize these movements as being essentially local in scope, too small and powerless to take on the forces of globalization.

London-based accelerationist thinkers Nick Srnicek and Alex Williams have in my view dismissively termed these forms of local resistance "folk politics."[109] One wonders if degrowth, too, is a form of "folk politics" to them.

The question, though, is how the "luxury communism" proposed by Bastani avoids the pitfall of "folk politics." The answer Bastani provides is elections. He is a champion of electoralism, calling for the advancement of "left-wing populism."[110]

His argument goes like this. The state should use policy to promote progress if it means bringing about the advent of the coming technological revolution even slightly more quickly, which will only help instantiate the economy of abundance earlier. To this end, governments should proactively fund research and development through foundations and subsidies. This would involve further deregulation at a grand scale as well. When parties consciously incorporating such measures into their platforms naturally arise, the masses will respond with electoral support. This is the Bastani-style strategy for realizing left-wing populism.

But even as Bastani aims to effect a major transformation in society, his vision of bringing about a communist revolution through elections seems, to borrow the phrase used by accelerationists to criticize environmental movements, rather too "folk" a version of politics itself—"folk" in a different way, a dangerous way.

First of all, it feels like a "folk" belief to think that the kind of transformation of the relations of production necessary to overcome capitalism could be achieved through political reform. It's a belief that amounts to nothing more than a form of politicalism.[III]

THE WAGES OF POLITICALISM— WILL GOING TO THE POLLS CHANGE THE WORLD?

Politicalism is the belief that if we simply select good leaders within a framework of representative democracy, we can leave it up to these politicians and experts to put optimal policies and laws in place for us. The hope is for charismatic leaders to appear, and when they do, we vote them into power. The key to reform thus lies in electoral behavior. However, this has the effect of narrowing the field of political action to elections. It is reduced to an image war played out in the media and on social networks, based on the publicization of manifestos and the selection of candidates.

The costs are clear. Bastani is calling for communism. Communism is, by definition, an enormous transformation of the relations of production. But because Bastani's version of communism is an essentially *political* project effected through politics and policy, my view is that it loses sight of the aspect of the transformation that must take place in the field of production—that is, it loses sight of class struggle.

"Old-fashioned" forms of "excessive" direct action and class struggle such as strikes, demonstrations, and sit-ins are seen as liabilities in an electoral context, bad for a candidate's image and damaging to what should be a "united front," and so they become strategies jettisoned by politicalism. Furthermore, the "folk"

opinions of the common citizen become suppressed by the author-
ity of experts. The top-down reforms instituted by the politicalists
may seem efficient at first glance, but the wages of this efficiency
can be seen in the narrowing of the field of democratic participation and the damage done to the sense of agency felt by those who
do participate.

Indeed, a social revolution effected through politics and policy
is favored by economists like Joseph Stiglitz. Recall Žižek's critique
of Stiglitz. Representative democracy cannot expand the purview
of democracy itself and cannot effect a revolution across all of
society. Electoral politics always reaches its limit when faced with
the power of capital. Politics does not exist separately from the
economy—rather, it is subordinate to it.

A nation cannot pass a law powerful enough to overcome the
power of capital—if this were possible, surely we would have done
so by now. This is why the field of political possibility must be
expanded through a social movement that confronts capital directly.

REVITALIZING DEMOCRACY THROUGH CITIZENS' ASSEMBLIES

One example of such a movement is the phenomenon of the "climate
citizens' assembly" that has been getting traction lately. Citizens'
assemblies became well known due to the efforts of the British
environmental movement Extinction Rebellion and the French
Mouvement des Gilets Jaunes (Yellow Vest Movement). Though the
backgrounds of these movements differ greatly, they both have
used the obstruction of roads, bridges, and traffic in general to par-
alyse cities and disrupt the daily lives of average citizens.

These sorts of "excessive" actions, performed without apparent
care for possible arrest, have attracted worldwide attention. The
Yellow Vest Movement is frequently understood as a movement of
working-class people, including truck drivers and farmers, rebelling
against the higher fossil fuel taxes introduced by the "elitist" French
president Emmanuel Macron as a measure to fight climate change.

But these "Yellow Vests" were actually joined by many who were calling for tougher measures to fight climate change as well. They criticized Macron for raising the fossil fuel tax while lowering overall taxes on the very richest, who, after all, are responsible for such a disproportionate amount of total carbon dioxide emissions. Further, he reduced public transportation services in rural areas, forcing more people into using personal vehicles as their primary mode of transportation.

Facing such overwhelming criticism, Macron announced that he would hold a *Grand Débat National* in January 2019. This great debate ended up consisting of more than 10,000 meetings held by local governments all over the nation, resulting in the presentation of more than 1,600 proposals. But many still felt that this was a debate in name only, and dissatisfaction and unrest persisted until Macron agreed to hold a long-promised Climate Citizens' Assembly in April of the same year.

In this way, France witnessed the inauguration of citizens' assemblies numbering about 150 participants each. These assemblies were entrusted with the creation of policies meant to reduce greenhouse gas emissions by 40 percent (compared to levels in 1990) by the year 2030.

A unique feature of a citizens' assembly is the method by which members are chosen. Rather than elections, a lottery system is used. This is a decisive difference from a parliament or congress made up of elected officials. Of course, this lottery system isn't completely random—the makeup of each assembly is meant to reflect the makeup of its community as closely as possible in terms of age, gender identity, educational background, residency, and so on. Experts give lectures addressed to these assemblies, after which debates take place between members, and in the end, consensus across the membership is measured by vote.[112]

One result of this process that deserves particular attention took place on June 21, 2020, when a French citizens' assembly presented its resolutions to then-minister of ecology Elisabeth Borne. This assembly, made up of 150 citizens chosen by lottery, proposed approximately 150 measures for stopping climate change. It included a proposal to ban the construction of new airfields starting in 2025, the complete stoppage of domestic air routes, the banning of car advertisements, the lowering of speed limits on expressways, and the introduction of taxes on the wealthy to fund further climate change prevention measures. On top of this, the assembly proposed the introduction of a specific clause related to climate change into the French constitution as well as the introduction of a referendum that would define and enforce the crime of "ecocide."

The radicality of these citizens' assembly proposals cannot be separated from the radically different mode of democracy the assemblies exemplify. And it is equally important to note that this radically different mode of democracy only came about as a consequence of a social movement.

Movements like the Yellow Vests and Extinction Rebellion are frequently criticized for not offering concrete demands. Yet when their demands for a more democratic mode of political participation are actually met in the form of citizens' assemblies, the policies proposed end up being quite concrete indeed.

If these movements had simply made concrete demands in the first place, they might have affected public policy in some form or other, but the current system of representative democracy would have remained unchanged. Furthermore, the content of the proposals would surely not have been so revolutionary. What the creation of the citizens' assemblies shows us is that social movements can renovate democratic processes and use the power of the state without sliding into "climate Maoism."

WE, WHOM SUBSUMPTION BY CAPITAL HAS RENDERED POWERLESS

We may see the possibility for changing the nature of politics in examples like the citizens' assembly, but many may still find arguments like Bastani's more attractive. It's simply easier to decide to leave it up to the political elite and experts to determine the future. If Bastani is correct, then all we need to do is talk to our friends on social media and watch movies on Netflix, and as long as we remember to vote, society will transform into one in which we no longer have to worry about student loans, uncertain employment, or the effects of climate change.

I cannot find in Bastani's proposals any call for radical change to the Imperial Mode of Living. As long as we vote, we can keep buying new iPhones every two years, wearing the fast fashion available at Zara and H&M, and eating burgers at McDonald's. To take this argument to its logical extreme, Bastani's luxury communism would permit the freedom to circle the globe in a private jet just to sample the food at every five-star restaurant on Earth. After all, new technology will allow us to ignore the limitations of both dwindling resources and the Earth's environment.

As we can see, Bastani's version of luxury communism could slide very easily into simple consumerist abundance, paving the way for recapture by capital. In other words, while Bastani's argument may seem radical at first glance, in my view it's really nothing more than a rebranded version of Silicon Valley–style capitalism. He criticizes capitalism but is actually quite enamored of it. And some have become very enamored of Bastani's version of accelerationism.

Underpinning the appeal of this vision are the unprecedented levels of powerlessness experienced by those of us living in the Global North. We feel, unconsciously, that we have no say and that we cannot exist without capitalism. This leads to an impoverishment in the imagination of the Left, which should be thinking up solutions to this conundrum. Humanity has created technology that allows it to dominate nature more than ever before, and its influence is felt on every inch of the planet. Yet we feel more powerless than ever in the face of nature's power.

People who are highly conscious of environmental issues are not exempt from this. We may be intentional about buying healthy, natural products that are certified organic, but we still end up buying these products, along with our salmon and chicken, in markets or having them delivered to our homes wrapped in pretty plastic packaging.

Most of us lack the ability to raise animals or catch fish for ourselves and prepare them properly for consumption. In the past, not only could people do these things, they could even make the tools necessary to do it themselves. Compared to them, we have been swallowed up by capitalism completely, lacking the power to support ourselves as living beings. We cannot survive without commodities; we have lost the know-how necessary to live in concert with nature. All we know how to do anymore is live our urban lifestyles supported by the exploitation of the periphery.[113]

The briefly fashionable Lifestyles of Health and Sustainability (LOHAS) movement attempted to address this without facing the problem of powerlessness within capitalism, focusing instead on attaining sustainability through consumer choices, and it ended in failure. Ultimately, changes at the level of consumer consciousness were always swallowed up by capital due to our enmeshment in a commodity-based economy that always strives for increased growth.

Marx termed capital's ability to swallow us like this "subsumption." Our lifestyles are subsumed by capital, rendering us powerless. The theoretical limit of Bastani's vision is, in short, the same as LOHAS's: its inability to overcome subsumption by capital.

FROM SUBSUMPTION TO HEGEMONY

The complete subsumption by capital has robbed us of our skills and self-sufficiency. We are unable to survive independently of the power of commodities and money. We've become so used to the ease of this way of life that we've lost the ability to imagine any other.

To borrow a phrase from the American Marxist Harry Braverman, the subsumption of all of society by capital has resulted in a fundamental separation of "conception" from "execution."[114] Let me briefly explain what he means by this.

Human labor was once united at the level of conception and execution. For example, artisans conceived of chairs they wished to create, then used planes and chisels to make them. A unified flow linked all parts of the labor process.

Capital, though, views this unified flow as inconvenient. If production depends on the skills and discernment of an artisan, it becomes hard to control the pace of production and labor, which makes it difficult to raise productivity. In the worst-case scenario, forcing things might end up injuring the artisan's pride and leading him or her to walk off the job.

Therefore, capitalists carefully observed the work performed by these artisans. They defined each component part of the process, measured the minimum time needed to perform each task, and then recreated the process in a more efficient way by dividing the labor among several people. This rendered the artisan obsolete. Now each component task in the creation of a product could be done by anyone, and the production was faster than if they were done by a single artisan, potentially at the same level of quality or higher.

This leads to the fall of the artisan. Capital ends up monopolizing the power of conception. The workers employed to replace the artisans simply execute the commands of capital. Thus conception is separated from execution.

This rationalization of work resulted in the sharp rise in productivity for society overall. Yet the productive skills of individuals went down. Modern workers are unable to create a complete product alone in the manner of the artisans who came before them. The people who assemble our computers and televisions are ignorant of how computers and televisions actually work.

This means that workers cannot conceive of their own labor outside its performance under capitalism. Having lost their self-sufficiency in this way, workers become mere cogs in a machine. They've lost the subjective agency of conception.

The subsumption of capital in the contemporary moment has moved beyond the labor process into all areas of life. This means that no matter how much our productivity increases, we remain unable to conceive of a future. Rather, we find ourselves ever more obedient to capital as it infiltrates our lives, unable to do anything but execute what it commands.

TECHNOLOGY AND POWER

It's when the hegemony of capital completes the process of subsumption that the true danger of Bastani-style accelerationism becomes clear. If the only objective is to accelerate the development of new technologies, the gulf separating conception and execution will only grow more profound and the hegemony of capital will only grow stronger.

This will lead to a situation in which decisions about what technology should be used and how become the sole province of the handful of experts and politicians who hold the power to conceptualize the future. All capital would have to do then is subsume them, too. Even if various problems really can be solved using new technology, it's very likely that its application will be handled in a top-down manner by a select, elite few.

Let's think about this issue using the example of a climate change countermeasure that has received a great deal of attention: geoengineering.

There are many types of geoengineering, but one thing they all have in common is their attempt to control the climate by intervening into the planetary system itself. These various forms of intervention include cooling the Earth by blocking sunlight with sulphuric acid aerosols sprayed into the stratosphere, placing a mirror in space that would deflect the sun's rays, and promoting photosynthesis by scattering iron in the oceans and thus making it fertile enough to support the growth of huge amounts of phytoplankton. Even Paul Crutzen, who coined the term Anthropocene, has been supportive of investigating geoengineering, making it one of the most symbolic projects undertaken in the context of this new era.

But there are many unknowns that remain regarding what effects releasing huge amounts of iron or sulphuric acid into the environment would have on the climate and oceans, including possible side effects on the Earth's ecology and people's lives and livelihoods. It's quite likely, for example, that acid rain and air pollution would increase, which would negatively affect farming and fishing. If rainfall patterns changed significantly, many regions may find their situations becoming even worse than they already are.

There's every reason to believe that such costs are being carefully calculated to hit Asia and Africa more than America or Europe. It's the same capitalist story all over again: the burdens are externalized as the metabolic rift grows deeper.

And yet, even in the face of this possibility, some still persist in calling for a top-down society ruled by capitalists and a handful of politicians.

ANDRÉ GORZ ON TECHNOLOGY

Readers hearing these criticisms of accelerationism may well respond by accusing me of denying the productive force and technological advancements fostered by capitalism and expecting everyone to go back to nature to live primitive, rustic lives. But Marx didn't argue for turning our backs on scientific progress and returning to the tired traditions of the archaic agricultural commune.

As we saw in the previous chapter, it's true that late in his life, Marx rejected the progressive view of history and praised the steady-state economy enforced by the traditions of precapitalist communal societies. But this does not mean he rejected technology and modern science outright. Marx never stopped calling for producers to use the natural sciences to develop a "rational way" to regulate the metabolic relation with nature.[115]

It's a fallacy to approach this problem as a binary between embracing science and rejecting it. It's clear that we will have to continue developing renewable energy, energy-saving technologies, and communications technologies beyond their current capacities.

It may behoove us here to turn our attention to the nuanced arguments offered by the French Marxist André Gorz. Toward the end of his life, Gorz was clear in his warnings about the dangers of technological development under capitalism. According to him, the productivism that leaves everything up to specialists is, in the end, linked to the negation of democracy, "a negation of both politics and modernity at once."[116]

Gorz asserts that to avoid the dangers of productivism, it's important to distinguish between "open technologies" and "locking technologies." Open technologies are those that involve exchange with others, that relate to communication and cooperative industry. By contrast, locking technologies are those that divide people, that turn users into slaves and monopolize the provision of products and services.[117]

A prime example of a locking technology is nuclear power. Nuclear energy was lauded for a long time as "green energy." But ostensibly for reasons of security, nuclear energy is isolated from the general public, and information about it is kept classified and tightly controlled. Such a situation leads to cover-ups, inviting major accidents to occur. It's impossible to manage nuclear energy production democratically. It is in this sense a locking technology, unadaptable to democratic management and necessitating the top-down politics of centralized control. In this way, technology and politics cannot be separated. Specific technologies demand specific forms of politics. In the context of climate change, technologies like geoengineering and NETs are democracy-negating, locking technologies *par excellence*.

LOCKING TECHNOLOGIES ARE INADEQUATE FOR GLOBAL DANGERS

Geoengineering will cause large-scale, irreversible transformations across the entire Earth. This is why, before we rush headlong into depending on geoengineering to save us from the mess we've created by pursuing economic growth at all costs, we should stop and ask ourselves, aren't there more democratic ways to address the problem?

The more pressing the danger becomes, the more people's priorities will shift to simple survival as the room to stop and reflect disappears. Once that happens, it will be too late. We may even let an iron-fisted leader drastically curtail our freedoms if embracing autocracy means our lives will be saved. Awaiting us at the end of such a scenario is an antidemocratic authoritarianism founded on chauvinistic nationalist sentiment: in short, "climate Maoism."

The climate crisis is a global crisis. It's no longer the kind of thing that can be externalized onto a periphery, and the Global North will ultimately be unable to evade its destructive consequences. What we are facing is a test to see if all of humanity can work in solidarity to prevent the worst-case scenario from coming to pass.

It is precisely because we are being tested this way at this moment that we cannot turn to locking technologies like geoengineering and NETs that prioritize the welfare of those living in the developed world at the expense of those living in the designated exterior.

TECHNOLOGY ROBS US OF OUR IMAGINATIONS

Furthermore, the problems with technology are deeply rooted. All across the world, the discovery of new technologies is ballyhooed as leading us into new, previously unimaginable futures. It has reached the point that some are calling this moment a technological "revolution." Increasing amounts of tax money and labor power are being poured into the development of these "useful" technologies, while budgets for the "useless" humanities are being cut more and more.

Yet what future are ecomodernist technologies like geoengineering and NETs, which seem so marvelous at first blush, locking us into? Isn't it a future of the same old lifestyle supported by burning fossil fuels? The marvelousness of these dream technologies resides in their support of the status quo, and therein lies their ineffectiveness, their true inadequacy for the crisis. These possible futures divert our attention from what can be done now—what *must* be done now. In this sense, technology itself becomes ideological, concealing the irrationality of the present system.

To phrase it another way, technology ends up suppressing and pushing out the possibility of creating a completely different mode of living in the face of the impending crisis, of forming a society truly independent of fossil fuels.

The present crisis should be spurring us to reflect upon our behavior and imagine a truly different future for ourselves. But the imaginative and conceptual power necessary to do so has been taken from us by the experts and specialists who have monopolized technological development. There are surely too many who still think that we can sit back and let technology solve climate change for us.

In short, the ideology of technology has become a cause of the widespread impoverishment of imagination in today's society. To regain the ability to imagine a different society, we must take our creative power back from its subsumption by capital. The degrowth communism first hinted at by Marx is one such precious well from which this creative power can spring.

IMAGINING A DIFFERENT ABUNDANCE

Why have I spent so much time critiquing Bastani's brand of left-accelerationism? Because understanding the problems with his argument makes the task before us easier to accomplish. In short, we must overcome locking technologies and domination by huge corporations like Google, Apple, Facebook, and Amazon (GAFA) in order to wrest our powers of imagination back and find a new path toward a new future.

The first step toward doing this is developing *open* technologies. We must resist the seductive power of the top-down politics produced by locking technologies and foster the development of technologies that enable people to manage their lives themselves. The price mechanism of the market is based on scarcity, and thus it can be disrupted by abundance.

But if we truly wish to challenge capitalism, we must redefine abundance in such a way that it cannot be confused with capitalistic consumerism. We should stop betting our future on the possibility that exponential growth in technological development will take care of things for us, exempting us from the need to modify our mode of living. Rather, we must change our mode of living so we can discover new forms of abundance. In short, we must break the link between economic growth and abundance and think seriously about how abundance can be linked to *degrowth*.

We must face reality in our call for a new abundance. If we do, we'll notice something right away. The world undergoes "structural reforms" over and over to foster growth, and yet the results are always the same: gaps widen between the rich and the poor, and rates of both poverty and austerity increase. The wealth held by the twenty-six richest capitalists in the world is equivalent to the total assets belonging to the world's poorest 3.8 billion people, nearly half the world's population.[118]

Can this be a coincidence? Surely not. We usually think of capitalism as something that provides wealth and abundance, but the truth is quite the opposite. Capitalism is a system that functions by producing scarcity.

Scarcity and abundance. In the next chapter, we will think along with Marx about the relationship between the two to capitalism as we reflect more deeply on the role of capital in the Anthropocene.

6

Capitalism's Scarcity, Communism's Abundance

CAPITALISM PRODUCES SCARCITY

Which produces plenty, capitalism or communism? Most people would surely answer capitalism. Capitalism has fostered technological advances previously unseen in human history and brought about a society rich in material objects. This is what most people think, and there's some truth to this view.

But it's not the whole truth. We must interrogate the issue. Doesn't capitalism in fact cause *scarcity*, at least for 99 percent of us? Couldn't we say that the more capitalism advances, the more economic hardship does?

A classic example of how capitalism produces scarcity is land. Looking at places like London and New York, we can see how property values have made even small apartments cost, in some cases, upward of millions of dollars. Rents have climbed to thousands of dollars a month, with some rising as high as tens of thousands a month. These properties are frequently bought and sold not as places to live but as assets for speculation, and as this speculation increases, so does the number of apartments where no one actually lives.

As this process unfolds, longtime residents who can no longer afford rising rents are driven from their homes, some joining the city's unhoused population. From the point of view of social justice, it's a scandal that so many people are experiencing homelessness in a city filled with vacant apartments bought and sold as mere speculation.

It's extremely difficult even for members of the comparatively affluent middle class to live in Manhattan. They have to work themselves nearly to death just to make rent. Furthermore, it's nearly impossible for independent business owners to open offices or storefronts in the downtowns of cities like London and New York. These opportunities are only affordable and available to large-cap corporations.

Is this state of affairs one of plenty? For most people, it resembles nothing but scarcity. And indeed, capitalism is a system that ceaselessly produces scarcity.

On the other hand, communism, contrary to popular belief, aims to provide a certain kind of abundance. For example, if land speculation were banned, prices would surely fall by half or even two-thirds, wouldn't they? Land prices are created artificially. A drop in price does not affect the land's use-value at all. Yet people are willing to work themselves to the bone just to barely afford access to this land. The use-value becomes the entire recompense for the price—the tiniest morsel of "abundance."

To explicate the relationship between the scarcity produced by capitalism and the abundance provided by communism, it's helpful, naturally enough, to turn to Marx. The concept of "primitive accumulation" found in *Capital*'s first volume provides a particularly interesting insight into the issue.

PRIMITIVE ACCUMULATION CREATES ARTIFICIAL SCARCITY

Generally speaking, "primitive accumulation" refers to the "enclosure" of land that occurred in England between the sixteenth and eighteenth centuries. It is the process by which farmers were forcibly expelled from farmland that had previously been managed communally.

Why does capital compel enclosure? The answer is simple: profit. Enclosure was instituted to allow farmland to be used as pasture for more-profitable sheep, or, in the case of the Duke of Norfolk, to enable the consolidation of land into large estates so that higher concentrations of capital could be accumulated through their management.

Driven violently from their land and deprived of both their homes and means of production, the farmers streamed into cities seeking work.[119] Enclosure is what paved the way for capitalism to really take off.

Marx's study of primitive accumulation is commonly understood as a form of bloody prehistory to the rise of capitalism, illustrated by vivid historical descriptions. But this is an inadequate way to understand Marx's intended argument, which uses primitive accumulation to critique capitalism itself.

The fact is, Marx's theory of primitive accumulation provides the tools to grasp the process of enclosure from the perspective of scarcity and abundance. According to Marx, primitive accumulation is the precise process by which the abundance of the commons is divided up to create more and more artificial scarcity. In short, capitalism has fed itself by impoverishing people, starting with this inaugural moment of enclosure and continuing through today.

Let's go back in history and examine exactly how this came about.

DIVIDING THE COMMONS MADE CAPITALISM TAKE FLIGHT

During our previous discussion of the archaic *Markgenossenschaft* of the Germanic peoples and the *mir* of Russia, we saw how precapitalist communal societies lived and worked while managing their land cooperatively. Even after these communes were broken up by war and the advance of market-based societies, communal land practices persisted in the form of commons and public farmland.

As a primitive means of production, land was managed by all of society rather than as private property to be bought and sold freely by individuals. In England, cooperatively managed land came to be called commons. People used these commons as needed to support themselves, gathering fruit, firewood, mushrooms, and the like, as well as using these areas for fishing and hunting game birds. It's said they also gathered acorns to feed livestock.

But such commons cannot coexist with capitalism. If everyone already has what they need to support their lifestyles, commodities won't sell in the marketplace. No one would make purchases if they don't have to. So the commons ended up thoroughly divided via the process of enclosure, converted entirely into exclusionary private property.

The result is what I call "the tragedy of the commodity," in opposition to Hardin's famous "tragedy of the commons." This resonates with what Marx elucidated in *Capital* as the prehistory of capitalist formation in England, which is known as the "primitive accumulation of capital." People were driven from the land where they had made their lives and were deprived of the means to support themselves. To rub salt in the wound, activities like hunting and gathering were redefined as crimes like trespassing and theft. As a result of the loss of communal management, the land became arid, both crops and livestock falling into decline, and fresh vegetables and meat became impossible to find.

Having lost their livelihoods, many of these former farmers ended up moving into cities, forced to become wage workers. Because their wages were so low, they couldn't afford to send their children to school, as everyone in the family needed to earn as much money as possible just to eke out a living. Yet meat and vegetables were still too expensive to buy. The quality of their diets became low, the variety of foods restricted. Without time or money, traditional recipes became useless, and they were reduced to surviving on boiled or fried potatoes. The quality of their lives had clearly fallen.

But things looked different from the perspective of capital. Capitalism is a social arrangement whereby anyone can buy and sell anything freely in the marketplace. These people who'd lost their land and means of survival now had to sell all they had left— their labor power—for money, which they then had to spend in the marketplace to regain their means of survival. This is the process by which a commodity-based economy suddenly advances, as all the conditions are now in place for capitalism to take flight.

FROM THE COMMONS OF HYDROPOWER
TO THE MONOPOLY OF FOSSIL CAPITAL

This process occurs in many realms, not just with land. Capitalism's launch necessitated the separation of people from the commons of rivers and streams as well. Rivers and streams don't just provide drinking water and opportunities for fishing. Moving water is also a source of abundant, sustainable, and free energy.

The Industrial Revolution in England is inseparable from the use of coal, a fossil fuel, and thinking about how this history is bound up with the present-day climate change crisis, the availability of water power becomes quite interesting. Why did this free source of energy fall by the wayside? The answer seems bound up with the issue of scarcity we've been tracing. It appears that capitalism had to ignore an abundant and available energy source in favor of one that only existed in specific places—and that was therefore scarce and able to be monopolized—in order to thrive.

The work of Marxist historian Andreas Malm is very helpful for understanding this history, as he discusses it in detail in his book *Fossil Capital* (2016).[120] Malm explains why humanity turned away from hydropower, tying it to capitalism's emergence.

Histories of technological development are, generally speaking, frequently written with a "Malthusian" analysis underpinning them. They usually go something like this. A shortage occurs in the supply of a resource because of the expansion of the economy. Prices go up due to this shortage, which provides the incentive to discover or create a cheaper alternative. This is the typical Malthusian mode of explanation. But as we've seen, hydropower already naturally exists in abundance and is a perfectly sustainable and cheap source of energy. To borrow Gorz's term, it's an open technology, one that can be managed as a form of commons. So why the shift from free, abundant hydropower to costly, scarce coal? The typical Malthusian explanation fails to convince in this case.

Malm explains that this shift cannot be understood without factoring in the crucial role played by capital. At its origins, industry started using coal not just as a simple energy source but as a form of "fossil capital." Coal and oil, unlike the moving water of rivers and streams, can be transported and can also be monopolized. These natural attributes come to possess, for capital, social meanings.

The shift from waterwheels to steam-powered machinery meant that factories could easily be moved into cities, since they no longer had to be located beside rivers. Labor power was comparatively scarce in areas near rivers and streams, giving workers power over capital. But when factories moved into cities filled with masses of workers seeking jobs, the problem was solved, as now capital had the upper hand.

Capitalists were thus able to monopolize this scarce energy source in the context of the city and use it as a basis from which to organize production. Due to this, the power relations between capitalists and labor reversed completely. Coal is an exemplary form of locking technology.

This move resulted in the sustainable energy of hydropower being pushed aside. With coal as the dominant energy source, productivity rose, but the air in the cities became choked with pollution and the workers themselves were overworked to the point of death. And of course, the switch to coal put us on the path, from that point on, to ever-increasing carbon dioxide emissions from burning fossil fuels.

THE COMMONS WERE ABUNDANT

One thing that's important to remember is that before the emergence of primitive accumulation, the commons of land and water were plentiful and abundant. Any member of the communal societies organized around them could freely take what they needed and use it.

This is not to say that usage was completely unrestricted. There were specific social mores that had to be respected and sometimes punishments for those who transgressed them. But as long as these rules were upheld, the commons were a form of open, freely accessible, communally owned wealth.

Precisely because the commons were a commonly held form of wealth, people made use of them appropriately, coexisting with nature without intruding upon it unduly, since there was no profit-making incentive for overproduction. The sustainability of the *Markgenossenschaft* rests on this as well.

The private property system that followed the enclosure of the commons, by contrast, destroyed this sustainable relationship between humans and nature founded on abundance. Land that had previously been free for anyone to use became available only to those able to pay for the privilege (via rent). Primitive accumulation divided the commons, imposing artificial scarcity in their stead.

What was once the commons became private property. Once people spent money to possess a piece of land, they could do anything they wished with it—no one had the right to interfere. Everything became dependent on the discretion of the private property owner. Due to this form of freedom, no one can stop an owner from doing anything, even if it results in the worsening of the lives of people who live nearby, even if the land itself becomes depleted, even if the water in the area becomes polluted.

It is in this way that the quality of everyone else's lives suffered in the name of the rights of the few.

PRIVATE RICHES DIMINISH PUBLIC WEALTH

In fact, this contradiction was already being written about in the nineteenth century. James Maitland, the eighth Earl of Lauderdale, a Scottish politician and scientist active in the early 1800s, discussed this exact problem in his book, *An Inquiry into the Nature and Origin of Public Wealth* (1804).

Lauderdale's ideas became famous enough that the contradiction between the commons and private property is known even now as the Lauderdale Paradox. This paradox states that any increase in private riches comes about only through the diminishment of public wealth.[121]

For Lauderdale, "public wealth" refers to wealth accessible to the common man, defined as "all that man desires, as useful or delightful to him." By contrast, Lauderdale defines "private riches" as that which is accessible only to an individual. As he puts it, private riches "consist of all that man desires as useful and delightful to him; which exists in a degree of scarcity."[122]

In other words, the sole difference between private riches and public wealth is scarcity.

Public wealth is commonly held by citizens and therefore isn't defined by scarcity. Private riches, on the other hand, cannot increase without increasing scarcity—otherwise they don't exist. Private riches come into being through the deliberate creation of scarcity by dividing up the public wealth needed by everyone. Private riches only increase by increasing scarcity.

As hard as it might be to believe, what Lauderdale was seeing occur right before his eyes was the justification of sacrificing others in the name of personal enrichment. Which, indeed, is the essence of capitalism itself and is a problem that persists today.

For example, water is abundant and is something that everyone desires and needs. In such a situation, water should be free. It is thus an ideal form of public wealth. Yet these days, water has, by whatever means necessary, been rendered scarce, commodified, and assigned a price. And thus another freely available form of public wealth disappears. Enclosing water in plastic bottles and selling it for a profit increases private riches (and of course plastic waste) instead, which also increases "national wealth" as measured in monetary terms.

And indeed, we can see Lauderdale's book as a direct criticism of Adam Smith's proposition that the wealth of nations is measurable as the sum of the *private riches* held within them. Lauderdale asserts that while national wealth, when measured this way, increases as private riches increase, the true wealth of a nation resides in its citizens' access to public wealth—that is, the very commons whose *diminishment* produces private riches in the first place. This results in the citizens of a nation losing their rights to the things they need to live and falling into destitution. The nation's wealth defined in terms of money may increase, but the quality of its citizens' lives decreases. In this way, Lauderdale differs from Adam Smith, as he sees that the true wealth of a nation resides in its *public* wealth.

Lauderdale provides many examples of what he means. He points to capitalists burning tobacco harvests when there was a surplus and forbidding the cultivation of vineyards to reduce wine production, measures designed to create scarcity of tobacco and wine.[123] Large harvests should, by rights, be cause for rejoicing. Yet excessive supply lowers prices, so surplus ends up wasted as a form of price support.

Scarcity expands as abundance shrinks. This is the fundamental truth behind the Lauderdale Paradox's assertion that private riches only increase at the expense of public wealth.

THE OPPOSITION BETWEEN USE-VALUE AND VALUE

Lauderdale, though, stops at simply identifying the paradox that bears his name. Marx, by contrast, investigates this contradiction between riches and wealth as he develops his theory of the fundamental contradiction within the commodity form.

To use Marx's formulation, we should translate Lauderdale's "wealth" into "use-value." Use-value indicates the quality in something—for example, air or water—that satisfies a human need or desire. Use-value existed well before the advent of capitalism.

"Riches," by contrast, are measured in money. They're based on calculating the "value" of commodities and thus don't exist outside a market economy.

According to Marx, the logic by which value is conferred onto a commodity becomes dominant under capitalism. The ultimate objective of capitalist production is to increase this value.

As a result, use-value is reduced to simply a means of creating a commodity's value. Even though fulfilling people's needs through the production of use-value was the entire point of economic activity in precapitalist societies, it becomes completely displaced under capitalism. Indeed, use-value can end up sacrificed, even destroyed, in the name of driving up value. Marx refers to this as the "contrast or opposition" within a commodity between use-value and value as he criticizes the irrational nature of capitalism.

NOT THE TRAGEDY OF THE COMMONS, THE TRAGEDY OF THE COMMODITY

Let's think about water again as an example of what I mean. Water exists in abundance, at least in Japan and in many countries in the Global North. Water possesses unassailable use-value, as everyone needs it to live. For this reason, it should be freely accessible and belong to no one. But water has become a commodity circulating in plastic bottles. Becoming a commodity has transformed water into something scarce, unable to be used without spending money.

The same thing is happening to the water supply. In the hands of private companies, the provision of water must become a profit-seeking enterprise, meaning that water bills must rise to levels higher than what's necessary to support the system.

There are those who believe that assigning water value is a way to encourage people to use it wisely as a limited resource. If it were free, people would waste it. This is the famous "tragedy of the commons" popularized by American ecologist Garrett Hardin.

But assigning water value is simply a way of treating it like capital, leading to a slippery slope toward water becoming an object of investment to drive up that value as much as possible. Once this happens, problems emerge one after the other.

For example, water becomes unavailable to people living in poverty who are unable to pay their water bills. The companies managing the water supply are incentivized to intentionally lower the amount of water available so as to raise its value and increase their profits. The companies may save money on staffing and water management costs without regard to possible drops in water quality. In the end, the division of the commons of water results in the disruption of its universal accessibility, sustainability, and safety.

We can see here how the commodification of water raises its value. Yet this rise in value results in the worsening of people's quality of life and even its use-value. This is the ultimate result of transforming the formerly free and abundant resource of water into a scarce, costly commodity. It thus seems more accurate to speak not of the tragedy of the commons but the tragedy of the commodity.[124]

THE PROBLEM IS NOT JUST NEOLIBERALISM

British Marxist geographer David Harvey has defined primitive accumulation as "accumulation by dispossession," arguing that the capitalist class's use of the nation-state to extract wealth from the labor class is the essence of neoliberalism. He goes on to argue that Marx's definition of this accumulation by dispossession as capitalism's "original stage" is a "disadvantage" of his theory.[125]

But Harvey is missing a crucial part of Marx's argument about primitive accumulation. It's in fact Harvey who seems to be limiting the definition of dispossession to the workings of neoliberalism.

Marx in no way intends for his account of primitive accumulation to act merely as a prehistory of capitalism. Rather, Marx is pointing out that the artificial creation of scarcity by dividing the commons is what lies at the very heart of primitive accumulation. The development of capitalism is the extension, continuation, and expansion of this fundamental process of primitive accumulation.

The austerity measures characteristic of neoliberalism may end soon. But whether neoliberalism persists or fades away, primitive accumulation will continue for as long as capitalism does. Capital will always continue to profit from creating and expanding scarcity. For the 99 percent, this spells the infinite continuation of immiseration and impoverishment.

SCARCITY AND DISASTER CAPITALISM

Let me take this opportunity to sum up my argument thus far. The commons constitutes use-value for the average person. Precisely because something is useful and necessary to everyone, a communal society will designate it as a commons and ban its monopolization, managing it instead as a form of public wealth. Uncommodified, no value can be assigned to the commons. They are abundant and freely available to the people. It's a state of affairs that capitalism naturally finds intolerable.

Once some sort of method is found to make these commons artificially scarce, the market can assign them value in precisely the same way that land was made scarce by dividing it through enclosure. Once that occurs, a landowner is able to make money through rent.

The use-value of land and water remains unchanged before, during, and after the process of primitive accumulation. What changes as the commons are converted into private property is scarcity. The more scarce a resource becomes, the more its commodity value rises.

The result of this process is the impoverishment of the average person, as access to the necessities of life are cut off or rendered difficult to attain. Value, as measured by money, may rise, but people's lives become meager and poor. The average person's quality of life is intentionally sacrificed in the name of driving up value.

Indeed, even waste and destruction, as long as they serve to produce scarcity, are seen by capitalism as business opportunities. This is because waste and destruction transform abundant things into scarce ones, providing the opportunity to maximize value.

Climate change is, in this sense, a business opportunity like no other. Climate change renders water, farmland, and habitation scarce. As this scarcity rises, demand rises too, until it surpasses supply and provides a prime opportunity for capitalists to reap huge profits.

This is a classic example of taking advantage of a state of shock to make money—a shock doctrine for the age of climate change.

From a profit-seeking point of view, sacrificing people's lives and livelihoods is an extremely *rational* way to maintain scarcity. One has only to recall the classic disaster capitalism practiced by the superrich of the United States, whose "coronavirus shock doctrine" resulted in them increasing their riches by $565 billion during the first three months of the COVID-19 pandemic.[126]

The COVID-19 pandemic exploded due to the poor health system after long-lasting neoliberal reforms, which exacerbated the COVID-19 shock and took thousands of lives every day. But multiple stimulus checks were issued not to maximize the protection of people's health but rather to bail out industries that are harmful to the environment, such as the airline industry, the cruise industry, and the fossil fuel industry. At the same time, big tech companies such as Facebook, Google, and Amazon profited from the pandemic because the deregulation of digital technologies and data usage created new frontiers of capital accumulation, a situation that Naomi Klein has called the "Screen New Deal." Furthermore, the Trump administration rolled back more than a hundred environmental rules on US territory for an indefinite period. Such deregulation, which would have been unimaginable in normal times, facilitated capital accumulation at a huge cost to the planet.

Scarcity created through the sacrifice of use-value serves to increase personal riches. This is the opposition between use-value and value that shows us the irrationality at the heart of capitalism.[127]

MODERN WORKERS ARE WAGE SLAVES

Let's now take a closer look at the scarcity created through the dissolution of the commons.

Once deprived of the commons, we're thrown into a world of commodities. The first thing we face there is the scarcity of money. The world is overflowing with commodities. But without money, we cannot buy them. Money will allow us to buy anything, but the methods for attaining money are very restricted, leading to impoverishment. Therefore, we must chase money as hard as we can just to survive.

We once worked for a few hours a day and then, once our needs were met, spent the rest of the day at leisure. We napped, played, talked to each other.[128] These days, though, we are forced to work long hours at the behest of another just to receive a little money. Time has become money. Which means time has become scarce—we cannot afford to waste even a minute of it, not even a second.

Marx frequently referred to the conditions of labor under capitalism as slavery.[129] Workers are like slaves in the sense that they must work and work without breaks, irrespective of their will. Modern workers are infinitely replaceable under capitalism. Once fired, workers face starvation and even death if they cannot find new jobs. Marx called this form of precariousness "absolute poverty."[130] It's an expression that contains in concentrated form the essence of capitalism as a system that produces perpetual meagerness and scarcity. To translate it into the terms of the argument contained in this book, poverty is caused by absolute scarcity.

THE POWER OF DEBT

Capital employs another form of artificial scarcity to complete its dominion. This is monetary scarcity caused by debt. The process of consumption under capitalism is driven by the unlimited stimulation of want, which leads not to wealth but to taking on loans. Shouldering debt in this way locks workers into obedience, forcing them to become capitalism's pawns.

A prime example of this is mortgages. Home loans hold great power over people's lives because of their size. People who take on loans so large they take thirty years to pay off lock themselves into working even harder and for longer in order to keep up with their debt. Workers end up internalizing the capitalist work ethic to pay back their loans. They work longer and longer hours to rack up overtime pay, postpone retirement, and sacrifice family time just to receive raises.

In some cases, a couple's dual income is not enough, and each person ends up working two jobs—one during the day and one at night—to make ends meet. Either that, or they make do with less or worse food, getting by on fast food or other cheap, heavily processed foods. Before long, the situation deteriorates until they no longer know why they live the way they do. They bought their house to improve their lifestyle, but debt has made them into wage slaves, destroying everything good about their lives.

A worker's industriousness is, of course, very convenient for the capitalist system. On the other hand, long work hours lead to the overproduction of fundamentally useless things, which in turn leads to the destruction of the environment. Long work hours rob us of the ability to clean or repair our houses, making us even more dependent on cheap and easy-to-get commodities to maintain our lifestyles.

This is how capitalism advances—through the production of artificial scarcity. As long as the opposition between use-value and value persists, the economy can grow exponentially and its benefits will never reach all of society. In fact, people's lives will become even less enjoyable and satisfying. This is the reality that most of us already experience every day.

THE RELATIVE SCARCITY PRODUCED BY BRANDING AND ADVERTISEMENTS

There's a consumer dimension to scarcity's ability to lower our quality of life and leave us unsatisfied. If workers are driven to ceaseless labor, this creates huge volumes of commodities. So now people must be driven to ceaseless consumption as well.

One way to provoke ceaseless consumption is through branding. Advertisements confer special meaning onto logos and brand images, inducing people to buy things they don't need at prices higher than they're actually worth.[131]

The result is the creation, through branding, of an entirely new standard for discriminating between commodities that have identical use-value. Common goods become transformed into commodities possessing a unique allure. This is how scarcity is manufactured in a world overflowing with more essentially identical products than anyone could possibly need.

From the point of view of scarcity, branding can be seen as its "relative" form. It enables a type of differentiation that allows one consumer to attain a higher social status than another.

To put things more concretely, let's imagine that everyone drove a Ferrari and wore a Rolex. Those products would suddenly have the same value as a Subaru and a Swatch. The social status associated with a Ferrari resides solely in its scarcity. This makes it a status symbol. To put it the other way, the use-value of a Rolex and a Swatch are exactly the same from the point of view of their functionality as watches.

Relative scarcity is a struggle with no end. All we have to do is open Instagram to see someone who owns something that's "better" than what we own, and the nice things we do buy are superseded quickly by newer models that make what we have seem faded and outdated. Consumer dreams can never come true. Even our desires and feelings have been subsumed by capital, twisted into unattainable shapes.

We buy things ceaselessly to help realize our ideal versions of ourselves, to attain our dreams and desires, and this drives us to work even harder, only to end up consuming more. There's no natural end to this cycle. Consumerist society compels people into ceaseless consumption by making sure that the dreams promised by owning commodities will never come true. Scarcity, felt by the consumer as dissatisfaction, is a fundamental engine of capitalism. It guarantees that no one in its grip will ever truly be happy.

Furthermore, the cost of all this useless branding and commercialism is incredibly high. The marketing industry is the third biggest in the world after food and energy production. Between 20 percent and 40 percent of the price of commodities is their packaging, it's said, while in the case of cosmetics, the price of the packaging can rise to as much as three times that of the product itself. Of course, the alluring packaging uses a tremendous amount of plastic that ends up simply being thrown away.[132] All this happens even as the use-value of the products remains exactly the same.

Is there any way out of this vicious cycle based on scarcity? To resist the artificial scarcity of capitalism, we must create a new society based on abundance. The way out is the degrowth communism envisioned by Marx.

RECLAIMING THE COMMONS IS COMMUNISM

Marx states that communism is "the negation of negation."[133] The first negation is the division of the commons by capital. Communism, as the negation of this negation, aims to reclaim the commons and restore *radical abundance*. Capitalism manufactures artificial scarcity to perpetuate itself. This makes abundance its natural enemy.

The key to restoring radical abundance is the reclamation of the commons. Indeed, it's the commons that will enable us to overcome capitalism and restore radical abundance in the twenty-first century.

It may be useful here to explain in concrete terms how the commons relate to abundance. Access to electricity, like to water, is a human right that should never be denied or left up to the marketplace to provide. The marketplace denies access to electricity to those without sufficient funds.

This doesn't mean it should be nationalized. The nationalization of electricity does nothing to prevent the adoption of locking technologies like nuclear energy that sacrifice safety for utility. Moreover, fossil-fuel power plants are frequently placed in areas where poor people and minorities live, polluting their air and causing myriad health problems.

An alternative way to attain the goal of treating electricity as a form of commons is citizen management. This is a practice by which sustainable energy can be easily handled at the citizen level. A concrete example of this is the recent spread of renewable energy production by citizen-owned power plants and energy cooperatives. A twist on privatization, it's a form of *private citizen-ization*.

THE "PRIVATE CITIZEN-IZATION" OF THE COMMONS

The point here is that unlike nuclear and fossil-fuel power generation, solar and wind power don't require exclusive ownership to function. There is a radical abundance of solar and wind power. They are truly unlimited and free. Furthermore, unlike uranium and oil, they can be established relatively cheaply and managed by anyone, anywhere. To borrow Gorz's terminology, renewable energy is an open technology.

But such attributes are fatal to capitalism. If an energy source like solar power were truly decentralized and available to everyone, it would be impossible to monopolize and make scarce, which would make it very hard to monetize.

This creates a dilemma for capitalism. The inability to induce scarcity is the same as the inability to make money. This is one reason why industry has lagged so far behind in the conversion to sustainable energy within the market economy. We can see here the emergence of the opposition between the scarcity of capitalism and the abundance of the commons.

This is why the private citizen-ization—that is, the citizen management or municipalization of energy production—is essential to the widespread adoption of renewable energy. It represents an opportunity to construct small-scale power networks amenable to democratic management, free from the profit motive and capitalizing on its inherently dispersed and decentralized nature. If we miss this opportunity, it will be all too easy for capital to reestablish its monopoly on the resource by clearing forests to create mega solar power plants, leading to further desertification and environmental destruction.

This sort of private citizen-ization has already been attempted with some success in the UK, Denmark, and Germany. The nonprofit citizen management of renewable power production has also sprung up in Japan. After the disaster at the Fukushima Daiichi Nuclear Power Plant in March 2011, there have been an increasing number of initiatives encouraging the local production of energy for local consumption, including solar panels being placed in abandoned fields, citizen pressure on city councils to encourage investment in private and green stocks, and so on.[134]

When energy is produced and consumed locally, the money paid for electricity stays in the community. Since the energy company is nonprofit, the money can also be reinvested into improving the lives of people who live in the area. In the process, citizens gain a stronger sense of sharing a commons that improves their lives, leading to more active participation in civic life and resource management.

Once such a cycle begins, a region's environment, economy, and society begin to synergize, revitalizing the community. This is what it means to transition to a sustainable economy via the commons.

WORKERS' CO-OPS—RETURNING THE MEANS OF PRODUCTION TO THE COMMONS

The commons consist of more than just power and water. The means of production must be returned to the commons as well. I'm talking here about workers' cooperatives—organizations allowing workers to invest jointly in the co-ownership and comanagement of the means of production without interference from capitalists and shareholders.

Workers' co-ops play a crucial role in workers regaining their autonomy and power of self-determination. Every member of a co-op invests in, owns, and operates the enterprise. It's the workers who have the agency to discuss and determine what sort of work will be done and how it will take place.

What enables such agency is that the co-op is not the private holding of a CEO or a group of shareholders, nor a national enterprise run by a government, but a form of *socialized ownership* by the workers themselves.

This form of worker-ownership has a long tradition behind it, starting in 1844 with the opening of the Rochdale Society of Equitable Pioneers. Marx himself praised the efforts of workers' co-ops, "acknowledg[ing] the co-operative movement as one of the transforming forces of the present society based upon class antagonism."[135] He asserts that the workers' cooperative movement shows how it was possible to replace the current capitalist system founded on scarcity with "the republican and beneficent system of the association of free and equal producers, even calling the workers' co-op an example of 'possible communism.'"[136] The workers' co-op in German is called *Genossenschaft*, and Marx frequently used the adjectival form of that word—*genossenschaftlich*—synonymously with "association."

Why would he do this? The answer is related to the expropriation of farmers' means of production during primitive accumulation, enclosing the commons to create scarcity. Workers' cooperatives return the means of production to the hands of the producers through solidarity between workers and thus help restore radical abundance.

DEMOCRATIZING THE ECONOMY WITH WORKERS' CO-OPS

Interestingly, workers' co-ops and the socialized ownership of enterprise have experienced a recent revival of popularity within the British Labour Party.[137] This has mostly stemmed from their role as alternatives to the disappearing welfare state.

The welfare state was the prevalent model of wealth redistribution in the twentieth century, one that refrained from touching relations of production directly. Put broadly, it was an attempt to give the profits taken by companies back to the rest of society through income and corporate taxes.

Behind this effort lay the subsumption of labor unions by capital in the name of raising productivity. Unions decided to cooperate with capitalists due to the belief that it would increase the "pie" available for redistribution. The price of this cooperation was the weakening of workers' autonomy.

In contrast to labor unions subsumed by capitalists, workers' co-ops aim to transform the relations of production themselves. Introducing democracy into the workplace allows workers to suppress competition amongst themselves and make joint decisions about development, education, and restructuring on their own terms. While co-ops do still support the continuation of their enterprise through profit making, they are not at the mercy of market-driven speculation by investors or the drive to maximize short-term profits above all else.

A major strength of co-ops is that workers can work as they wish. Co-ops aim to advance a "social and solidarity economy" (SSE) that restores the regional community through workplace training and management practices. Through labor, workers can make investments weighted toward the long-term prosperity of the region. This amounts to nothing less than the democratization of the economy by making the realm of production itself into a form of commons.

This might sound like a far-fetched dream to some. But it doesn't have to be. Workers' cooperatives of this sort are spreading all over the world. Spain's famous Mondragon Corporation is a federation of workers' cooperatives with a long history, boasting more than 74,000 members. In Japan, too, there have been workers' co-ops in sectors like nursing, childcare, forest management, agriculture, waste disposal, and so on for close to forty years. Their collective reach amounts to more than 15,000 people.

Even in the capitalist stronghold of the United States, workers' co-ops have developed in remarkable ways. Notably, the Evergreen Cooperatives in Ohio, Cooperation Buffalo in New York, and Cooperation Jackson in Mississippi are examples of citizen movements to revitalize communities by addressing problems related to housing, green energy, food, waste disposal, and so on. Based on the successful model of Mondragon Corporation in the Basque Region of Spain, these co-ops aim to build a network of democratic institutions to empower workers. In particular, Cooperation Jackson, established in 2014, attempts to tackle the structural problems of class, gender, and race by addressing the needs of poor, unemployed, Black and Latinx female residents.

In economic systems that prioritize profit making, essential work like childcare, cleaning, cooking, and waiting tables pays very little. This means that this type of work is done primarily by women of color and immigrants and divided from the rest of society, resulting, in the end, in the worsening of work conditions. It is another vicious cycle.

To remedy this, workers' cooperatives aim to make essential work of this sort autonomous and desirable. The hope is that improving wages and working conditions will help revive the community by overcoming barriers of race, class, and gender.

Of course, as Marx also pointed out, even a workers' co-op will end up sucked back into capitalistic competition if it makes even one false step. Once this happens, cost-cutting and efficiency will once again rule the day as profit making becomes the most important thing once more. The system as a whole must change to prevent this. Yet from a "no one left behind" standpoint, workers' cooperatives can become the basis by which to transform society overall, premised as they are on resisting the impoverishment, discrimination, and inequality fostered by capitalism.

A RADICAL ABUNDANCE DISTINCT FROM THE GDP

Workers' co-ops and the private citizen-ization of power grids are just a few examples of possible actions. The possibility exists in many sectors—including education, healthcare, the internet, and the so-called "sharing economy"—for radical abundance to be returned to the people. For example, what if Uber were publicly owned, turning its platform into a commons? To use an example of a different sort, what if the vaccines and drug treatments for COVID-19 were a worldwide commons?

Via the commons, the collective management of productive activity can spread horizontally throughout society, independent of both the market and the state. The result would be goods and services becoming abundant that are now scarce and difficult to access without money. In short, the goal of restoring the commons is an attempt to reduce the reach of artificial scarcity and increase the radical abundance available outside the realms of consumerism and capitalism.

An important point to remember is that the management of the commons can easily occur independently of the state. Water can be managed by autonomous regional bodies, and electricity and farmland can be managed at the citizen level. Sharing-economy services can be managed collectively by app users. There are even cooperative platforms for advancing innovation in the IT sector. The space taken up by commodification decreases as radical abundance is restored. For this reason, the GDP would also decrease. This is degrowth.

But this does not mean people will become impoverished. Rather, as the space taken up by mutual aid, independent of the exchange of money, expands, people will be released more and more from the pressures of work. The amount of time regained by the average person just through this shift would be immense.

Once a more stable lifestyle is attained, the amount of time and effort we can devote to mutual aid will increase, as well as the capacity to devote ourselves to nonconsumerist activities. There will be more opportunities to do sports, go hiking, take up gardening, and get back in touch with nature. We will have time once again to play guitar, paint pictures, read. We can host those close to us in our homes and eat with friends and family. We will have the free time to volunteer or engage with politics. The consumption of fossil-fuel energy may decrease, but the community's social and cultural energy will rise up and up.

Compared to cramming ourselves into crowded subways every morning and eating our deli lunches in front of our computers as we work nonstop for hours and hours every day, this is clearly a richer lifestyle. We would no longer have to shop online or drink ourselves into oblivion just to rid ourselves of the stress of simply surviving. If we regain the time necessary to cook for ourselves and exercise, our health will surely improve as well.

We have been overworking ourselves with the belief that economic growth will bring prosperity and happiness. Capitalists benefit greatly from our diligence. Within the framework of capitalism, our dream may be to become rich, but the essence of capitalism is scarcity, making it impossible for this dream to be realized by everyone.

It's a system we must break away from and replace with degrowth. The only way to fully realize a system of radical abundance is through degrowth communism. People's lives will become stable and rich without relying on economic growth to provide for us.

With the closure of the enormous gulf separating the ultra-rich 1 percent from the 99 percent and the imperative of scarcity abolished, society will require fewer hours of labor to function. The lives of almost everyone will improve. Not to mention that the lessening of needless labor will, in the end, save the planet.

THE PLENTIFUL ECONOMY OF DEGROWTH COMMUNISM

This is a major paradigm shift. As we saw in chapter 3, degrowth has been criticized as a form of voluntary poverty up until now—a call for all of us to choose impoverished lives in the name of protecting the environment.

But this criticism is bound up in capitalist ideology—the "curse of growth." This ideology is rooted strongly enough that it bears repeating what's at stake here.

The system of austerity that demands we withstand impoverished, meager lifestyles is capitalism, as it's a system that depends on artificial scarcity to function. We're not impoverished because we don't produce enough; we're impoverished because scarcity is capitalism's essence. The basis for this is the opposition of use-value and value of commodities.

The austerity measures instituted by neoliberalism are perfectly fitted to capitalism in the sense that they increase artificial scarcity. By contrast, abundance demands a break from growth as a paradigm.

The economic anthropologist Jason Hickel, in his call for radical abundance, puts it this way: "While austerity calls for scarcity in order to generate more growth, degrowth calls for abundance *in order to render growth unnecessary.*"[138]

It's time to close the books on neoliberalism. Anti-austerity is what's needed now. But while simply spreading money around might combat neoliberalism, it will do little to put an end to capitalism. The only way to combat the artificial scarcity of capitalism is to restore the commons and reestablish radical abundance. This is the true anti-austerity measure—the move to degrowth communism.

GOOD AND BAD FREEDOM

We must put an end to capitalism and restore radical abundance to the world. What awaits us at the end of this struggle is freedom. It's a common misunderstanding of communism that it's a system that sacrifices freedom in the name of equality. The remainder of this chapter will therefore deal with the subject of freedom.

The radical abundance I've been arguing for up until now naturally compels us to redefine the concept of freedom. We must distinguish freedom from the American-style capitalist value system by which liberty is measured by the realization of a lifestyle that places the largest possible burden on the environment.

It's undeniable that humans are essentially free, and this freedom can be expressed by choosing to put ourselves on the road to extinction by destroying the very foundation of our lives. But this is not a good freedom—let's call this a bad freedom.

On this point, it's worth revisiting Marx's *Capital* and its important, if rather lengthy, passage on freedom:

> The realm of freedom really begins only where labour determined by necessity and external expediency ends; it lies by its very nature beyond the sphere of material production proper . . . Freedom, in this sphere, can consist only in this, that socialized

man, the associated producers, govern the human metabolism with nature in a rational way, bringing it under their collective control instead of being dominated by it as a blind power . . . But this always remains a realm of necessity. The true realm of freedom, the development of human powers as an end in itself, begins beyond it, though it can only flourish with this realm of necessity as its basis. The reduction of the working day is the basic prerequisite.[139]

Let's think about this argument's premise. Marx is distinguishing here between the "realm of freedom" and the "realm of necessity." The realm of necessity refers to the range of acts of production and consumption necessary to live. By contrast, the realm of freedom refers to the range of activities that may not be strictly necessary for survival are necessary to be *human*. These include the arts, culture, friendship, romance, and even things like sports.

Marx called for the expansion of this realm of freedom. In other words, the expansion of this category is the expansion of good freedom.

But this is not to say that this expansion should take precedence over the realm of necessity. People need food, clothing, and shelter to live, and we cannot do away with the forms of production necessary to provide these. The realm of freedom, he states, "can only flourish with this realm of necessity as its basis."

What we need to remember here is that the good freedom Marx wants to see flourish is not the material, individualistic consumerism of capitalism. Thanks to the workings of capital, our lifestyles may look rich at first glance. But driving this apparent prosperity is the endless need to satisfy material desires. The satisfaction of these desires—the all-you-can-eat buffets, the fast fashions discarded and bought anew every season, the meaningless branding—is all tied to the animalistic wants associated with the realm of necessity.

The realm of freedom, on the other hand, begins once we're *freed* from the clutches of such base desires. According to Marx, the true essence of human freedom lies in collective cultural activities.

This is why we must dismantle the system based on limitless growth that compels us to work long hours while driven to ceaseless consumerism; only then can we expand the realm of true freedom. This might result in lower levels of production overall than we have today, but the only way to realize happiness for all in a just and sustainable way is through the exercise of voluntary "self-limitation." The expansion of the realm of freedom will come not from the reckless pursuit of higher productivity but with the diminishing of the realm of necessity through restraint.[140]

WHAT THE NATURAL SCIENCES DON'T TEACH

The necessity of thinking of self-limitation as a form of good freedom is even greater now in the age of climate change. Thinking about the issue's relationship to the natural sciences will help make things clear.

At the beginning of this book, we saw how humanity is rapidly approaching a Great Divergence. It's a situation that demands that we seriously discuss what kind of world we want to live in and what sorts of choices we must make to bring that future world about. But the natural sciences cannot teach us what a society founded on fostering the realm of freedom will actually look like.

Science can tell us that the atmospheric density of carbon dioxide must remain below 450 ppm to stabilize global temperatures at 35.6°F above preindustrial levels. If we surpass this level, we can try employing technologies like geoengineering or bioenergy with carbon capture and storage to try to remove the carbon from the environment.

But what science cannot tell us is why a world with global temperature rises stabilized at 35.6°F is more desirable than one with global temperatures that have risen by 37.4°F. In other words, future generations will not have experienced life when temperatures were at their current levels and may even be happy enough living in a world where temperatures have risen by 37.4°F. Standards of satisfaction can be adjusted to given conditions, making them flexibly variable. This is the kind of argument someone like the economist Nordhaus, whom we discussed in chapter 1, might advance.

That's why it's up to us to take the utmost care in deciding at what temperature we wish to live and what sacrifices we're willing to make to achieve this. It's a question to be addressed democratically, not to be left up to scientists, economists, or artificial intelligence.

Put another way, we must realize that there's no naturally existing limit out there for us to surpass. Any such limit is determined by what sort of society we decide we want to live in—a question of social custom and habit. The setting of limits can only be done through political processes that follow from economic, social, and ethical determinations.

This is why we cannot be content to let these limits be set by experts and politicians alone. If we do, the shape of the world to come will be determined solely by their interests and worldviews, dressed up as scientific "objectivity." We have already seen how this works in the example of Nordhaus's biases toward promoting economic growth over stopping climate change being incorporated into the target values proposed by the Paris Agreement.

SELF-LIMITATION FOR THE FUTURE

Questions about what sort of world we want to live in must be debated passionately and democratically, reflecting as much as possible the voices of future generations.

This is especially true of climate change, which is irreversible. We cannot afford to try something else if our first attempt to combat it fails. In the realm of cloning and genetic editing, we have to be careful not to pass a certain limit beyond which irreversible changes to the very meaning of "human" may occur. In the same way, technologies like geoengineering have the potential to transform the meanings of "nature" and "planet" in ways that cannot be undone. These kinds of transformations would fundamentally undermine future generations' right to self-determination.

It thus becomes extremely important to prevent this situation by refraining from intervening too much. It is in this context that "self-limitation" arises as an increasingly important value.[141] Those of us living in the developed world must actively examine the things we produce, determining which ones are unnecessary and doing away with them, then looking at the things we do need and deciding how much of them it's really necessary to produce.

Under the hegemony of capital, which drives us toward unceasing, unrestrained consumption, it's nearly impossible to attain the freedom necessary to choose such self-limitation. After all, it's a condition of capital accumulation and economic growth that no one restrain themselves at all.

But let's think about it another way. If we were to voluntarily choose the path of self-limitation, this would be an anti-capitalist, revolutionary action.

Breaking away from unlimited economic growth and choosing a version of self-limitation that will enable everyone to thrive in a sustainable way will allow the realm of freedom to expand and realize a future of degrowth communism.

So, what concrete steps can we take to bring about such a future? This challenge will be addressed in the next chapter.

7

Degrowth Communism
Will Save the World

THE COVID-19 PANDEMIC IS ANOTHER
PRODUCT OF THE ANTHROPOCENE

Up until now, I have argued for the necessity of making a clean break from capitalism and transitioning to degrowth communism. Now I want to address the question of how to bring about degrowth communism in concrete terms, as well as how this transition will help solve the climate crisis.

But before I get to that, I'd like to briefly look at an example of a recent crisis caused by the Anthropocene: the COVID-19 pandemic. This "one-in-a-hundred-year" pandemic took the lives of a great many people and left behind an economic and social toll of historic proportions. Yet the worldwide damage wrought by climate change has the potential to dwarf it in magnitude. There's every possibility that future generations suffering from the effects of climate change will look back on the COVID-19 pandemic as a mere blip, just a temporary setback.

Despite this difference in scale, it's still instructive to look at COVID-19 as a precursor of what's to come. Both climate change and COVID-19 can be seen as examples of the contradictions lying at the heart of the Anthropocene. In that sense, they are both products of capitalism.

We've already seen how capitalism brought about climate change. It is, after all, the driving factor behind the expansion of economic growth and the environmental damage that goes with it into every corner of the globe.

The structural outline of the pandemic is similar. Capital has pushed as far as it can into the natural world to satisfy the ever-growing demands of the developed world, clearing forests to make room for large-scale agribusiness. Pushing into hitherto untouched parts of the natural world increases the chances of coming into contact with unknown viruses, but that's not all. The spaces cleared by human hands, unlike the complex natural ecosystems destroyed to make them, are monocultural spaces that lack the means to impede a novel virus's emergence. Evolving continuously, a virus can ride the flow of the globalized human population and spread throughout the world in the blink of an eye.

Moreover, the danger posed by a possible pandemic was well forecast by experts in much the same way that scientists have been passionately sounding the alarm about the dangers of climate change for decades.

The responses to the climate crisis and the pandemic resemble each other as well. When governments were faced with the decision to save lives or save the economy, they put off the fundamental issue in both cases, in the name of avoiding damaging the economy by doing "too much." Yet delaying effective countermeasures only caused even greater damage to the economy in the end, not to mention the grave loss of life.

DEMOCRACY SACRIFICED BY THE STATE

This is not to say that all countermeasures are good as long as they're early. The response of the Chinese government, which shut the country down during the first wave of COVID-19 in 2020, is an example of a top-down form of oppression enacted as an exercise of state power. It was a policy by which cities were locked down, people's movements were surveilled and regulated, and anyone who didn't follow the government's directives faced harsh punishments.

European Union countries laughed at these authoritarian measures at first, but when the pandemic began to spread within their borders, they instituted similar policies. Citizens by and large accepted these measures as unavoidable. South Korea sacrificed its citizens' personal privacy to suppress the virus's spread using digital technology. These examples speak volumes—as a crisis worsens, the state is called upon by specialists and experts to regulate the lives of citizens more and more invasively, a violation of individual freedom that citizens largely accept.

With this recent history in mind, I would like to return our attention to the four possible futures outlined in chapter 3. According to that broad schema, the strategies employed by Donald Trump in the United States and Jair Bolsonaro in Brazil are examples of the first future—a fascist style of governance. They prioritized economic activity above all else, replacing ministers and expert advisors who disagreed with them as they plunged forward. It was clear from their actions that they didn't mind if the only people who survived were those rich enough to afford costly medical care and those with jobs that allowed them to self-isolate and work from home. It was a situation in which the privileged received as many PCR tests as they needed while proclaiming that socially vulnerable groups, such as the poor, needed to take personal responsibility for their health.

Bolsonaro, for his part, noticed that the virus was spreading among the Indigenous peoples opposing the clearing of the Amazon rainforest and saw an opportunity to go against their wishes, lifting restrictions on logging in the area under the banner of economic recovery. It was a classic example of disaster capitalism.

The People's Republic of China and the EU, by contrast, took the health of their citizens seriously and instituted anti–COVID-19 policies in an exercise of state power. This corresponds to the third future outlined in chapter 3: an authoritarian, dictatorial form of governance. Prevention of the disease's spread became the alibi for restricting citizens' freedom of movement, freedom of assembly, and other freedoms with the power of the state.

The declaration of the state of emergency was capitalized upon by China as a way to crush the democratization movement in Hong Kong, while in Hungary, the dissemination of information about COVID-19 deemed "fake" by the government could result in up to five years in jail.

THE DEPENDENCE ON THE STATE CAUSED BY COMMODIFICATION

In any case, times of crisis create the strong possibility that the power of the state will display itself more and more flagrantly.

One reason for this is the transformation, since the 1980s, of various social relations into commodified forms under neoliberalism, whereby exchanges based on mutual aid are replaced by exchanges of money and goods. As we become accustomed to the situation, our ability to help each other, even the very impulse to do so, is uprooted. This means that when a crisis arises, we turn to the state rather than our neighbors for help. As the crisis deepens, we lose our ability to imagine how to rebuild our lives without increasingly invasive state intervention.

What will happen as people begin to demand increasingly heavy-handed interventions by the state to address the effects of climate change? Will this involve the building of walls, the expulsion of climate refugees, reliance on technologies like geoengineering to protect the few by sacrificing the many—in short, climate fascism? Or will it involve the total regulation of both industrial and individual carbon dioxide emissions by the state, including mechanisms of surveillance and punishment—in short, climate Maoism?

Whichever the case, the rise of politicians and technocrats will result in the sacrifice of democracy and human rights.

WHEN STATES CEASE TO FUNCTION

But we must be careful here. The argument I've been making is premised on the assumption that authoritarian governments are functional and effective. But when a crisis truly takes a turn for the worse, even such "strong" states may falter. Indeed, the COVID-19 crisis already showed how powerless most states become in the face of healthcare collapse and economic chaos. So it follows that authoritarian states may very well cease to function in the face of the climate crisis as well.

If this happens, the result is the second future mentioned in chapter 3—barbarism, a descent into a war of all against all.

This is no hyperbole. During the COVID-19 crisis, the extreme right-wing militia known as the Boogaloo movement attracted new recruits over social media as they planned an anti-government civil war in the United States.[142] In Michigan, armed citizens protesting lockdown measures marched on the state legislature, eventually even invading the state capitol building.

Moments of crisis like these show how vulnerable the Imperial Mode of Living really is. When the first wave of the COVID-19 pandemic hit, masks and hand sanitizer became impossible to find even in developed countries. This was the result of the pervasive outsourcing of production in the name of realizing the cheapest possible version of a lifestyle based on instant gratification.

Even though contagious viruses like SARS and MERS had spread in similar ways in the not-so-distant past, the major pharmaceutical companies of the developed world continued to concentrate their research and development on profitable medicines like antidepressants and treatments for erectile dysfunction, letting the development of antibiotics and antiviral medications lag far behind.[143] The cost of this choice was the collapse of resilience in most major cities of the developed world.

In the case of the climate crisis, food and water shortages will surely worsen. Countries where food self-sufficiency rates are low and overall resilience is weak will likely descend into panic. When this comes to pass, it won't be long before things devolve into a state of barbarism.

PRIORITIZING USE-VALUE OVER VALUE

In effect, these issues stem from the opposition between value and use-value as identified by Marx.

In the case of the pandemic, the use-value of medicine resides in its ability to cure ailments while the value resides in the price that can be charged for the medicine as a commodity. Between vaccines and erectile dysfunction drugs, the life-saving nature of vaccines makes them more useful. But capitalism prioritizes the ability to make money over saving lives. This makes medicines that will sell even at high prices more valuable than life-saving treatments.

Food is also looked at this way—the highest priority is given to products that can be sold expensively. But growing the highest-priced peaches or grapes for export does nothing to help overcome a food shortage at home. Capitalism's prioritization of a commodity's value even to the detriment of its use-value has created situations like this already all over the world. This is why we must break away from capitalism and transition to a society that prioritizes use-value again.

In chapter 3, I referred to the fourth possible future with the placeholder term X. Readers must have guessed by now what this X stands for: degrowth communism. This is the future to which we must aspire in order to save ourselves.

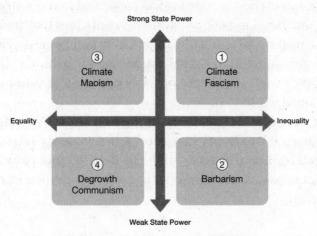

Figure 9. Four choices, four futures

COMMUNISM OR BARBARISM?

Why communism? Community self-governance and mutual aid are necessary to avoid a barbaric future ruled by far-right militias, radical groups like neo-Nazis, and organized crime syndicates like the Mafia. We must create democratic means to attain and distribute the essentials for living amongst ourselves, for ourselves. Which is why now, during a time of relative peace, we must start to nurture modes of autonomy and mutual aid to prepare for the coming crisis. One thing we learned during the COVID-19 pandemic was that we simply cannot rely on governments to provide this kind of help.

We can see how policies that stop at the point of turning away from excessive market fundamentalism and toward big government interventions into the marketplace are insufficient in the face of a crisis massive enough to shake the foundations of society itself. In other words, as we saw in chapter 2, green Keynesianism, which proposes large-scale fiscal stimulus plans that entail governments pouring capital into major industry, will never manage to actually reduce carbon dioxide emissions or keep the climate crisis from occurring. Moreover, we've also seen in chapter 3 how degrowth capitalism aiming to render Northern European–style welfare states sustainable is also inadequate for the coming crisis.

Inadequate half measures end up having no long-term effect on ongoing chronic emergencies. The fact is, social democracy in its present form has no real way to counteract the rise of rightist populism. We need to move away from the arguments of centrist leftism.

We must say it plainly—communism or barbarism! This is the only choice left! It's obvious we must choose communism. We must overcome our reflex to rely on experts and the state and proceed down the path to self-governance and mutual aid.

THOMAS PIKETTY'S CONVERSION TO SOCIALISM

This all might seem rather extreme. But it's a position taken, somewhat shockingly, even by superstar economist Thomas Piketty, author of the bestselling *Capital in the Twenty-First Century*.

Piketty is generally known as an economist of the liberal left who drew attention to the growing economic gap and proposed a strong scheme of progressive taxation as a solution. Piketty's compromise with capitalism was criticized, along with Stiglitz's, as "utopianism" by Žižek. And indeed, if limited to *Capital in the Twenty-First Century*, this criticism is valid.[144]

But Piketty's argument changes considerably in his subsequent *Capital and Ideology*, which came out in 2019. In this book, Piketty repeatedly speaks of "transcending" capitalism, proposing "*socialisme participatif*"—participatory socialism—in lieu of "domesticated capitalism" as the means to do so. Near the end of his book, Piketty declares, "The study of history has convinced me that it is possible to transcend today's capitalist system and to outline the contours of a new participatory socialism for the twenty-first century—a new universalist egalitarian perspective based on social ownership, education, and shared knowledge and power."[145] I can think of no clearer conversion to socialism in recent years.

Further, he observes with bitter irony that Social Democratic parties have turned their backs on the working class and become the province of wealthy intellectual elites, calling them the "Brahmin Left." He sharply criticizes this liberal left as having thus paved the way for the rise of the populist right, which claims to support the working class and the "common man." The left must once again remember to whose pain they are meant to attend. It is for this reason that Piketty takes up the banner of socialism.

THE CRUCIAL IMPORTANCE OF SELF-MANAGEMENT AND COMANAGEMENT

The content of Piketty's proposed solution is also worth examining. He still emphasizes, as he did before, the importance of property and inheritance taxes, but in the face of climate change, he also notes the limitations of carbon taxes imposed by the state. His point is that while market fundamentalism isn't the answer, neither is reliance on taxation by the state to solve the problem.

Confronting climate change allows Piketty to turn his attention to the site of production. What's essential, according to him, is the application of participatory socialism to production. Worker-led social ownership and participatory management must occur within the industry to realize this goal.

Piketty directs his criticism at the despotic decision making within a private corporation, which is dictated by the needs of a small number of major shareholders seeking to maximize their dividends. As an alternative, Piketty emphasizes the need for workers to take control of production either through self-management ("autogestion") or comanagement ("cogestion").[146]

In short, Piketty's response to the imminent climate crisis is to conclude that democracy cannot be sustained under capitalism. Simple redistribution of wealth and resources is therefore insufficient to preserve democracy; what is needed is socialism, which means that worker self-governance at the site of production is essential. In this, Piketty and I are in total agreement.

The phrase "participatory socialism" is important here as well. Self-management and comanagement, distinctive characteristics of his version of participatory socialism, are keywords corresponding to the main idea animating my own argument, the commons.[147]

The other thing Piketty emphasizes is the difference between participatory socialism and Soviet-style socialism. As a regime where all decision-making power resided with officials and experts who also controlled information and thought, the Soviet Union was antithetical to the democratic nature of participatory socialism.

In contrast to the authoritarian Soviet Union, participatory socialism is an attempt to transition to a sustainable society through nurturing the seeds of mutual aid and citizen self-rule. At this point, Piketty's position and that of the late Marx are closer than they have ever been.

HOW TO HEAL THE METABOLIC RIFT

Yet even Piketty stops short of explicitly endorsing degrowth. Even as he calls for participatory socialism, he envisions the transition as relying in large part on taxation—that is, on the power of the state. This is a problem. The more a state attempts to control capital through taxation, the more its power grows, sliding inexorably toward a form of state socialism I previously identified as climate Maoism. In this way, the system ends up moving away from Marx's degrowth communism.

I would like us to recall Marx's metabolic argument here. The production demanded by capital's quest to accumulate unlimited value leads to alienation from nature's fundamental cycles, which in the end results in an "irreparable rift" opening up between humans and nature.

According to Marx, the only way to heal this rift is to radically revolutionize the realm of labor so that production can be in sync with the cycles of nature again. As we saw in chapter 4, labor is what mediates the relationship between humans and nature. Marx's metabolic argument as developed in *Capital* asserts that humanity and nature are connected through labor. This is why it's crucial to change the nature of labor to save the environment.

To put it more provocatively, Marx believed that revolutionizing modes of distribution and consumption, transforming the political system, and changing people's values are all secondary concerns. Conventional wisdom misunderstands communism as consisting of banning private property and replacing it with nationalization, but the fact is that even modes of ownership aren't the real problem. The most crucial thing, above all else, is the revolutionary transformation of labor and production. This is the main difference between the form of degrowth proposed by this book and that of earlier degrowth theorists, who studiously avoided engaging with Marxism and workers' movements and thus failed to address the dimension of labor in their arguments.

Earlier degrowth theorists tended to focus on consumption instead, advising self-limitation to consumers—conserve water and electricity, stop eating meat, buy used goods, share products instead of buying them individually. But because they focused exclusively on changing people's values and on ownership and redistribution, they avoided dealing with the radical changes necessary in the realm of labor and thus never confronted capitalism directly.

There were thinkers in Marx's time as well, like French economist and philosopher Pierre-Joseph Proudhon, who wanted to bring about socialism through a revolution in the circulation of goods without touching modes of production. Marx criticized these thinkers, including Proudhon, severely for this. Only revolutionizing the site of production can empower us to transform the system as a whole.

REVOLUTION BEGINS AT THE SITE OF PRODUCTION AND LABOR

Emphasizing production this way may make my argument sound Marxist in a very old-fashioned way. But as I will explain in detail later, this book emphasizes production for different reasons than twentieth-century Marxism did. Furthermore, one of my goals is to reintroduce questions of labor into the degrowth and environmentalist movements that exclude them in favor of preoccupations with consumerism, enlightenment, and politicalism.

At the risk of repeating myself, I feel that reevaluating Marx's argument for revolutionizing labor is important precisely because it allows us to avoid the pessimism that's so easy to slide into in the face of a crisis as overwhelming as climate change.

The pessimism of our vision for the future in light of the climate crisis stems from the sheer size of the problem. There's nothing one person can do about it alone. Moreover, those who might possess such power—politicians, government officials, the business elite—seem to have no inclination to heed the voices calling for them to combat the crisis. It's therefore difficult to see any hope for change at the level of politics. Naturally, this leads to despair.

Yet if we allow ourselves to succumb to despair, all that awaits us is the descent into barbarism. There is one place left for people to take concrete action as directly concerned parties: the site of production. This is why the first step toward real revolution must take place there.

SMALL SEEDS OF CHANGE, SOWN IN DETROIT

There are small seeds planted in sites of production that are already starting to sprout. I would like to share a story here about one such instance. The setting is Detroit. A city that was once famously the center of American automobile production by companies like GM and Ford, Detroit became plagued with unemployment after that production moved away. Public finances collapsed, and after accumulating approximately $20 billion of debt, in 2013 the city went bankrupt. Over decades, Detroit had become a ruin, a symbol of the dashed dreams of capitalism.

People fled, law and order deteriorated, and the streets became wastelands. Yet those who remained never stopped trying to rebuild their city from the ground up.

As they did, they began to see their situation as an opportunity of sorts. The exodus of people and industry made land prices go down, which meant there was room to begin new projects. One of those projects was urban farming. Local volunteers and workers' co-ops led the effort to revitalize the urban wasteland through organic farming.[148]

This urban farming effort gradually turned parts of the city green. But even more importantly, the ties between members of the community, which had become so frayed and broken by the disorder that had reigned, were rebuilt. Local networks knitted themselves back together through the process of cultivating crops, selling them at local farmers' markets, and distributing them to local restaurants. Access to fresh vegetables naturally began to improve locals' health as well.

These kinds of movements are spreading all over the world. For example, Copenhagen has decided to plant fruit trees throughout the city that citizens are allowed to pick from for free.[149] The plan is to transform the entire urban area into an "edible city." We might call it a modern reinstatement of the commons. Radical abundance, incompatible with the logic of capitalism, begins like this.

Cultivating fruits and vegetables in the streets not only helps combat hunger by making food freely available but also fosters awareness of the dynamics of farming and nature in local citizens. Furthermore, no one wants to eat fruit infused with exhaust fumes. This leads to movements to expand bike paths in an effort to cut down on air pollution. Such efforts are first steps toward resisting car culture and reclaiming the abundance of roadways as a form of commons as well.[150]

In this way, people's imaginations of what is possible gradually expand, the progress that has already been made allowing for the conceptualization of hitherto unthinkable possibilities for the future. *What if Detroit's food supply were produced and consumed entirely locally? What if personal car use were banned within Copenhagen's city limits?* These concrete "what if" propositions are the means to overcome a poverty of imagination that simply accepts the status quo as unchangeable, cracking open a rift in the dominion of capital.[151]

Marxist critic Fredric Jameson once famously said it was "easier to imagine the end of the world than to imagine the end of capitalism."[152] But these seeds sown at the very site of production are starting to bear fruit, offering hope that could never be cultivated at the site of consumption.

SOCIAL MOVEMENTS ALLOW US TO TRANSCEND THE IMPERIAL MODE OF LIVING

The site of production is the site of community formation. As its boundaries expand, this community has the power to make a greater and greater impact on society as a whole. Movements arising out of labor hold the power to eventually transform even the realm of politics.

This is why the main issue this book addresses is not the dimension of lifestyle—not the Imperial Mode of Living—but rather the mode of production that makes this form of consumption possible.

In other words, it's imperative that we *transcend* the Imperial Mode of Living. The only way for such transcendence to take place is to overcome the system of production underpinning it.

But as I've said before, a "politicalist" model that attempts to solve the problem in a top-down manner will never work.

Which is not to say politics are irrelevant—some top-down measures are necessary if we are to adequately address the climate crisis in time. But any politics that attempt to take on climate change must also challenge the power of capital. The immense power of social movements will be indispensable in the effort to bring about such politics.

As Spanish sociologist Manuel Castells astutely put it, "Without social movements, no challenge will emerge from civil society able to shake the institutions of the state through which norms are enforced, values preached, and property preserved."[153]

If all we do is wait, the politics capable of combating the crisis of the Anthropocene will never arrive. But there's no need to wait. We can move forward right now.

CAPITAL IN THE ANTHROPOCENE

So, what exactly should we be doing? I would like to offer my answer to this burning question at last.

To review, Marx's *Capital* states that the only way to heal the metabolic rift between humanity and nature is to radically revolutionize labor so that production can be conducted in sync with nature once again. Labor mediates the relationship between humans and nature and thus constitutes the connection between the two. This is why it's crucial to change the nature of labor itself if we want to survive the coming climate crisis.

Just stating this, however, fails to explain exactly how changes in production and labor would actually help solve climate change. Did Marx think about how labor conducted under communism would help heal the "irreparable rift" in the metabolism connecting humanity to nature? A direct answer can't be found just by reading

Capital. Furthermore, there are scholars who have criticized Marx for the pessimism of his theorization of the metabolic rift.

The key to solving this conundrum is, once again, the perspective Marx arrived at late in life. After *Capital*'s initial publication, Marx continued his study of the natural sciences in search of ways to heal that rift. It's only by taking Marx's later perspective into account that we can reread *Capital* and divine how degrowth communism would be able to repair the metabolic rift.

Marxists of the twentieth century, by contrast, didn't take into account Marx's shift in thought at the end of his life and proceeded optimistically, sure that natural limits could easily be overcome once workers began to use technology and science on their own behalf. The idea was that technology would serve as the means of healing the metabolic rift.

But this sort of productivism proved to be mistaken as well as being incompatible with Marx's later thought. Traditional Marxism thus gave birth to many things, including chimeras like Bastani's blending of Marxism with Silicon Valley capitalism, but never to the form of communism Marx actually wished to bring about.

This is why it's so important to reread *Capital*, written as it was by Marx when he had yet to free himself from the curse of the progressive view of history, from the perspective of degrowth capitalism. In chapter 4, I assert that the true shape of his thought only emerges from *Capital* once we grasp the implications of his late-in-life study of ecology and communal social forms. And it is only this fully developed version of his ideas that can serve as the weapon we need to survive the present.

The true form of Marx's thought can be organized into five main points. These points are: 1) the transition to an economy based on use-value; 2) the shortening of work hours; 3) the abolition of the uniform division of labor; 4) the democratization of the production process; and finally, 5) the prioritization of essential work.

These demands might at first blush resemble those of traditional Marxists. But it should become clear soon enough that the ultimate goal here is quite different.

Marx's thoughts on degrowth have been overlooked for close to 150 years. This is why even demands that seem the same as those of earlier Marxists end up working out differently, as they've never before been formulated in a context of economic slowdown. In this sense, it is an update of *Capital* for the Anthropocene.

Degrowth communism can only lead to a transition to a sustainable economy insofar as it slows economic growth. Deceleration is the natural enemy of capitalism, which can only function by accelerating. It's impossible to sync production to the cycles of nature under a capitalist system that constantly demands unlimited profit. This is why the truly revolutionary movement isn't acceleration, it's *deceleration*.

Now, let's look at what we can do to bring about degrowth communism at last.

THE FIRST PILLAR OF DEGROWTH COMMUNISM: TRANSITIONING TO A USE-VALUE-BASED ECONOMY

Traditional Marxism, like degrowth communism, calls for the transition to an economy that prioritizes use-value by putting an end to mass production and consumption. It's even written out in explicit terms in *Capital*. So let's start there as we explore the principles of degrowth communism.

As we've seen, Marx distinguished between use-value and value in his analysis of the commodity. In chapter 6, we examined how capitalism's need for constant growth and capital accumulation makes a commodity's value more important than its use-value. Capitalism's number-one priority is increasing value. In its extreme form, this means that anything has value as long as it sells. Its utility (its use-value), its quality, its impact on the environment—none of that matters. It also doesn't matter if it ends up discarded immediately once it's sold.

Yet this increase in production for the purpose of increasing value gives rise to myriad contradictions if viewed from a wider perspective. For example, while lower prices enabled by mechanization stimulate demand and lead to massive amounts of a commodity being sold, this process also ends up destroying the environment.

Furthermore, increased productivity naturally leads to more things being produced, but capitalism's fixation on a commodity's value over its use-value gives rise to a system under which products are created primarily based on whether they sell rather than their utility for social reproduction. Indeed, the things truly necessary for social reproduction end up devalued.

As we saw at the outbreak of the pandemic, the way the production of necessary items to treat and prevent the spread of disease—ventilators, masks, hand sanitizer—was organized proved to be woefully insufficient. The factories making such items had been moved overseas as a cost-cutting measure, which meant that supposedly "developed" countries found themselves unable to produce enough masks for their citizens. This was entirely the result of sacrificing use-value in the name of capital increasing its own value. The net effect of this was the loss of resilience in the face of crisis.

Production focusing on status symbols and luxury goods, on advertising and branding, while ignoring use-value spells death in an age of climate crisis. There are so many vital issues to be addressed—guaranteeing universal access to food, water, electricity, shelter, and transportation; combating rising tides and flooding; protecting ecology; and so on. We must prioritize the production of things necessary to respond to the crisis, not things whose worth resides only in their capacity to produce value.

Communism would thus introduce a major shift in the purpose of production. Production would no longer be organized around creating value but rather around producing use-value as determined through social planning. Put a different way, fulfilling people's basic needs would be prioritized over increasing the GDP. This is the grounding principle of degrowth.

Marx would clearly be appalled at our current consumerism whereby productivity rises and rises to fulfill even people's most fleeting desires, endlessly. We must break ourselves out of our addiction to our present consumption practices and shift the emphasis of production to those things necessary for us to thrive while also practicing self-restraint. This is the form of communism necessitated by the Anthropocene.

THE SECOND PILLAR OF DEGROWTH COMMUNISM: SHORTENING WORK HOURS

We must shorten our work hours and improve the quality of our lives.

Shifting to an economy based on use-value will transform the dynamics of production in major ways. Meaningless work performed just to make money will decrease dramatically. Moreover, productivity will be consciously redistributed to focus on producing only those things truly necessary for the reproduction of society.

For example, marketing, advertising, and packaging whose only purpose is to needlessly stimulate people's desire for things they don't need would be banned. Consultancy and investment banking would no longer be necessary and would disappear. We would no longer need convenience stores and family restaurants to be open all night. Same-day delivery and overnight shipping would become things we can do without.

Once we stop producing so many things we don't need, we can reduce the hours people work across society as a whole. The reduction of meaningless labor that would result from shortening work hours would foster true prosperity. The shortening of work hours would also have positive effects on both people's lifestyles and the environment. Marx, in *Capital*, asserts that shortening work hours is a "fundamental condition" for transitioning to an economy based on use-value.

Today's society has surely attained a sufficient level of productivity already. Automation has enabled the unprecedented expansion of our productive power; at this point, there's really no reason why people shouldn't already have been liberated from wage slavery.

Under capitalism, though, automation means not the liberation from labor but the threat of replacement by robots—the threat of unemployment. Haunted by the specter of losing our jobs, we work ourselves to the point of death. Here we see the irrationality at the heart of capitalism. We must free ourselves of this irrationality as quickly as possible.

Through work-sharing, communism would aim to raise the quality of life (QOL) that cannot be expressed by the GDP.[154] Shortening work hours will reduce stress and make it easier for households to juggle childcare and caregiving.

It's important to note that shortening work hours is not a way to secretly raise productivity. It's true that "liberation from work" and "the fifteen-hour workweek" are catchphrases beloved not only by accelerationists like Bastani but some believers in degrowth as well. And indeed, the prospect of a "fully automated economy" sounds alluring. But Marx would surely add the following caveat: it's wrong to imagine the total elimination of labor as the endpoint of the reduction of work hours through automation, since liberation from work achieved while still raising productivity would only result in the further destruction of the planet.

Moreover, the shortening of work hours through automation needs to be thought of from another perspective as well: that of energy consumption.

Let's say a new piece of technology is introduced into a factory that allows a task that used to be performed by ten people to be performed by one. The factory would thus become ten times more productive, but that doesn't mean the individual worker will gain ten times the skill. Rather, the productive force of nine workers has been replaced by fossil fuel energy. The worker as wage slave has been replaced by fossil fuel as energy slave.

The important thing to consider here is the high Energy Returned on Energy Invested (ERoEI) of fossil fuel use. The return on investment of energy is the measure of how many units of energy are gained per unit of energy spent.

Looking at crude oil use in the 1930s, we see that for every unit of energy used, one hundred units of energy were gained. In other words, ninety-nine units of energy became available for us to use however we wished. Subsequently, though, the ERoEI of crude oil declined steadily until one unit of energy now yields a mere ten units of energy in return, a problem that has gained attention in recent years. The reason for this decline is the exhaustion of easily accessible sites of oil extraction.

The present ERoEI of crude oil is comparable to that of solar energy. Compared to the ERoEI of ethanol derived from corn, though, it's incredibly high. Ethanol's ERoEI, by which one unit of energy is produced for every unit spent, renders it meaningless as an energy source.

The transition to a decarbonized society demands that we stop using fossil fuels, despite their high ERoEI levels, and replace them with sustainable energy and biomass.[155] But even if we develop the means to fuel our cars and machines with sources of sustainable energy, it will be difficult to replace the fertilizers needed for agriculture, the concrete needed for construction, the iron and steel needed for manufacturing, and so on. The economy will inevitably slow down, impeding growth. The decrease in productivity associated with carbon dioxide emission reduction has been called the "emissions trap."[156] Reducing the amount of energy slaves without slowing production would necessitate humans working long hours again. This means that shortening work hours is inextricably tied to deceleration.

The slowing of production to reduce carbon dioxide emissions is something we have no choice but to accept. Because the emissions trap inevitably reduces productivity levels, it's even more important for meaningless work that creates little use-value to be reduced so that we can concentrate our productive forces on areas where they're truly needed. The "elimination of work" and "liberation from labor" promised by increased productivity cannot be realized in a decarbonized society.

This is why we must pay attention once more to Marx's call for the content of labor to become fulfilling and attractive again. It's only based on such a transformation that we can imagine the rest of the way forward.

THE THIRD PILLAR OF DEGROWTH COMMUNISM: ABOLISHING THE UNIFORM DIVISION OF LABOR

We must make work creative again by abolishing the division of labor that produces uniformity.

The images of life in the Soviet Union are so striking that it's shocking to hear that Marx called for making labor "attractive work." Even if work hours were to be shortened, people would turn to consumerism for stress release if the content of the work was still tedious and taxing. It's vital that not just the length but the content of work be transformed to reduce our stress and make our lifestyles more humane again.

Looking at the sites of production today, we can see how the subsumption by capital enabled by automation is making workers mere monitors of machines. The thorough standardization of production has made efficiency rise by leaps and bounds, but at the same time, it has stripped workers of their autonomy. More and more work has become more and more repetitive, meaningless, and dissatisfying to perform.

The previous generation of degrowth theorists evades the question of labor and thus never adequately addresses this issue. Their framework for degrowth aims to foster creative and social activities during the time spent *outside* work. To achieve this, they see automation as a way to reduce work hours as much as possible so that even if labor is tedious or taxing, it will be easier to bear.

Marx, by contrast, never wanted to eliminate or evade work. Rather, he envisions labor becoming "attractive work" that has "created the subjective and objective conditions for itself" to be the basis of an "individual's self-realization."[157] The point is not to simply increase the amount of free time that exists outside labor but to eliminate torment and meaninglessness from labor itself. We must transform work into something creative, an avenue for self-realization.

According to Marx, the first step toward returning creativity and autonomy to work is the abolition of the division of labor. Under the division of labor compelled by capitalism, work is restricted to its most standardized, efficient form. To make work attractive again, we must establish sites of production that allow workers to engage in a wide variety of tasks and activities.

This is why Marx calls on us repeatedly to overcome the opposition between mental and physical labor and the opposition between the urban and the rural to bring about the society of the future.

Marx emphasizes this point even in a late-period work like the *Critique of the Gotha Program.* He states that future society will come about once the "enslaving subordination of the individual to the division of labor, and therewith also the antithesis between mental and physical labor, has vanished," which means that work can become "not only a means of life but life's prime want." Only then can the "all-round development of the individual" be realized.[158]

To bring this about, Marx envisioned lifelong, equalized vocational education for all. This would allow workers to overcome subsumption by capital and truly become masters of industry. If we were to identify the surviving traces of such an impulse now, it would be in the intent behind vocational training promoted by workers' cooperatives and other types of cooperative organizations.

Here, too, the degrowth stance of Marx in his later years comes into play to push things further. Abolishing the uniform division of labor to restore humanity to labor necessarily entails deprioritizing the type of efficiency meant to promote economic growth. Work would be performed for the satisfaction of doing it or for its benefit to others, not to make money. The breadth of work would also expand, with workers performing more different kinds of labor, even rotating jobs in an equal manner or contributing on a local basis, all of which would naturally contribute to the slowdown of economic activity. We must see this as a good thing.

There is no need here to reject science or technology. Technology can, in fact, enable workers to engage in an even wider range of activities than they could previously. This is what it means for technology to be open, to use Gorz's terminology.

To develop technology in such a manner demands that we break away from an economy based on locking technologies that pave the way for dominion over workers and consumers—economies prioritizing profit over all else—and transition to an economy that prioritizes the production of use-value instead.

THE FOURTH PILLAR OF DEGROWTH COMMUNISM: DEMOCRATIZING THE PRODUCTION PROCESS

Democratizing the production process will decelerate the economy.

We must introduce open technologies to shorten work hours while maintaining the prioritization of use-value. Bringing about this revolution in work necessitates giving workers decision-making power over production. This is what Piketty calls "social ownership."

Social ownership would allow the means of production to be managed democratically as a form of commons. Decisions about what technologies to introduce into the workplace and how to use these technologies would be made democratically, through discussion and debate.

This principle applies to more than just technology. If decisions regarding energy and raw materials were made democratically, a great many things could change. For example, contracts with electric companies using nuclear energy could be cancelled in favor of choosing sustainable energy produced locally for local consumption.

The important thing, from Marx's point of view later in life, is that democratizing the production process would go hand in hand with decelerating the economy. Democratizing the production process means the communal management of the means of production through free association—in other words, deciding democratically what and how much should be produced, and how. Opinions on these matters will naturally differ sometimes, and consensus building in an environment free of authoritarian force takes time. One of the decisive changes social ownership brings about is a deceleration of decision making.

This differs greatly from the present system by which the decision-making process within industry reflects, above all else, the wishes of a small group of majority shareholders. The reason why major industry can respond quickly and decisively to moment-by-moment changes in the business environment is the undemocratic nature of the decision-making process, which is based solely on the wishes of management. This is what Marx calls the "tyranny of capital." By contrast, the form of production based on free association that Marx proposed would slow down economic activity due to its prioritization of democracy in the production process. The Soviet Union failed to adopt this principle, leading it to become a dictatorship of bureaucracy.

The democratization of the production process brought about by degrowth communism will transform production at all levels of society. For example, intellectual property rights and monopolized platforms, which allow new technologies to be patented and thus produce huge profits for certain sectors of major industry like pharmaceutical companies or the GAFA constellation, would be abolished. Knowledge and information would instead be treated as commons. We must restore the radical abundance of know-how. In so doing, it's very possible that the absence of competition for market shares and profit as motivations for development will decelerate the rate of innovation by private enterprise.

But this is not necessarily a bad thing. The development of locking technologies by capitalism to produce artificial scarcity can even obstruct scientific and technological progress. As Marx states in the *Critique of the Gotha Program*, people's abilities will flourish once they are freed from the tyranny of the marketplace, allowing for new innovations to occur that will help increase productivity through improved efficiency.

Communism aims to foster the development of new open technologies as commons enabling the kind treatment of both workers and the planet.

THE FIFTH PILLAR OF DEGROWTH COMMUNISM: PRIORITIZING ESSENTIAL WORK

We must convert to a use-value-based economy that prioritizes labor-intensive essential work.

As we saw in chapter 4, Marx, at the end of his life, broke away from productivism and accepted the importance of respecting natural limits. Building on this point, I would like to conclude by emphasizing the clear limits of the much-ballyhooed recent developments in automation and artificial intelligence.

Forms of work that are impossible to mechanize and thus require human labor are generally known as "labor-intensive industries." Care work is a classic example of such an industry. Degrowth communism will transform our society into one that values these labor-intensive industries. This transformation, too, will result in the deceleration of the economy.

To understand how prioritizing labor-intensive industries will slow down the economy, I would like to dig into the example of care work for a moment.

First of all, it is self-evident that automating care work is a difficult proposition. In realms of social reproduction that depend on care and communication, attempts to make work more uniform and standardized are doomed to fail due to the work's demands being complex, wide-ranging, and regularly involving irregular elements. These irregular elements cannot be done away with, making them difficult for automated systems to address.

This is precisely what makes care work a form of production that prioritizes use-value. A care worker cannot simply follow a manual to adequately help a patient eat, dress, or take a bath. Care work involves being flexible, responding on a case-by-case basis to another person's character and background. It involves listening to another's daily troubles, building a trust relationship while paying careful attention to small changes in another's body and soul. Childcare and education demand similar forms of attention.

These distinctive features make care work a type of "emotional labor." Unlike working on an assembly line, emotional labor becomes meaningless if performed without acknowledging the feelings of others. For this reason, it cannot be made more efficient and thus more "productive" by, for example, increasing the number of patients per care worker by two or three times. Care and communication are tasks that take time, and those receiving this care don't wish for it to be sped up.

Of course, nursing and caregiving can be streamlined into processes dictated by preset patterns and thus be made more "efficient" to a certain extent. But if the work is made too productive in this way in order to increase its value, the service's quality—that is, its use-value—will inevitably suffer.

Because of the difficulty of mechanization, labor-intensive industries like care work are seen as high-cost, low-productivity sectors of the economy. For this reason, unreasonable levels of efficiency are demanded of it by all levels of management, from the very top to the administration on the ground, and it is regularly subjected to irrational restructuring and cost-cutting measures.

BULLSHIT JOBS VS. ESSENTIAL WORK

The pressure capitalist society places upon essential workers is rooted in the extreme estrangement of value from use-value.

The highest-paying jobs right now are in industries like marketing, advertising, consulting, finance, and insurance, which makes these industries appear to be very important despite being almost entirely inessential to the reproduction of society.

As David Graeber has pointed out, even the people doing these jobs feel like society would hardly be affected if they stopped existing entirely. In short, the world is overflowing with meaningless work—with "bullshit jobs."

And indeed, we all seem to be endlessly attending useless meetings, preparing materials for presentations that no one cares about, writing business PR articles to post on Facebook that no one will read, and touching up pictures in Photoshop that will be appreciated by no one.

This contradiction is characteristic of a present moment in which people are flocking to industries that pay well but produce almost no use-value. Essential work, on the other hand, is low-paying and chronically plagued by worker shortages.

This is why we must transition to a society that prioritizes use-value. Such a society will necessarily place a high value on essential work.

This will be good for the environment as well. Care work is not only important socially, but it also results in few carbon emissions and uses few resources. Once we abandon our desire to constantly grow the economy, we can abandon our patriarchal fixation on manufacturing as well and head down the path of revaluing labor-intensive industries like care work. Such a transition will lead to a style of work appropriate to an age of lowering the energy return on energy invested.

This, too, will cause deceleration. After all, it's impossible to raise the "productivity" of care work without lowering its quality.

THE REVOLT OF THE CARING CLASSES

I've focused my discussion of degrowth communism on care work not just because it's an industry that's kind to the environment. It's also because people engaged in care work have risen up to resist the logic of capitalism all over the world. This is what Graeber has called the "revolt of the caring classes."[159]

Essential workers like care workers are currently forced to work long hours for low wages precisely because their work is helpful and considered to be a calling. Indeed, it's the fact that it's a calling that's being exploited. Furthermore, they are often held in contempt by administrators who create needless levels of supervision and regulation without actually helping anyone at all.

But these essential workers are beginning to rise up in protest. They've reached the limit of their ability to put up with how bad their working conditions have become. Moreover, cost-cutting has led to a worsening in the services they offer beyond the point they can stand as well.

The result is that even in Japan, situations like the mass resignation of childcare workers, formal objections filed from the sites of medical care, teacher strikes, and nursing strikes are becoming more and more prevalent. Beyond these examples, there have been work stoppages at 24/7 convenience stores and strikes among workers at expressway service areas. These were publicized on social media and gathered a great deal of support from the general public.[160]

This is part of a worldwide trend. Can these movements become more wide-reaching and connect to more radical movements elsewhere? Will we allow ourselves to act in solidarity with them when that happens? Or will we cling to our current bullshit economy that prioritizes meaningless work while denigrating work with actual use-value?

We have a choice between strengthening mutual aid and deepening the divisions between us. If we pick the right path to follow, it could lead to the creation of a new society, a remade community founded on democratic forms of mutual aid.

IMPLEMENTING SELF-MANAGEMENT

It's worth paying attention to the way the revolt of the caring classes can lead to the implementation of forms of self-management rather than being merely transitory forms of resistance.

One such possibility emerged in the US and the UK during the pandemic as social strikes broke out in various care sectors such as nursing, teaching, and caregiving.

The irresponsible management of schools and hospitals has long been a social problem due to neoliberal reforms. Austerity measures had already resulted in low salaries, long working hours, and chronic shortages of workers as well as the deterioration of facilities. The situation worsened because of the extra tasks necessitated by the pandemic.

From the point of view of both care workers and the patients/schoolchildren, this situation is wholly irrational. Here workers did not simply demand higher wages. Rather, the key strategy was bargaining for the common good. For example, teachers not only demand higher wages but also higher-quality education, including a cap on the number of students per teacher, more special education, more bilingual education, a nurse and librarian in every school, and protection for students from immigration enforcement authorities. During the strikes, schools offered packed lunches to children eligible for free school meals, and free parents took care of kids whose parents were at work.

Public servants working in the welfare sector, such as nurses and caregivers, are also striking for higher wages and more staff, as well as for more hospital beds in the national healthcare budget. In January 2023, more than seven thousand nurses in New York went on a three-day strike to protest the fact that there were too few nurses. Short-staffed hospitals are not just bad in terms of working conditions, stress, and burnout—crowded conditions put patients at risk.

In attempting self-management, the workers were clearly rejecting the exploitation and degradation of their calling by neoliberal reforms.

For these reasons, this is a prime example of an active "revolt" that returned production to the hands of the workers while maintaining the quality of the service being produced. Furthermore, solidarity between childcare workers (labor) and parents and patients (consumers) opens up the possibility of a more stable, cooperative form of self-management.

DEGROWTH COMMUNISM WILL HEAL THE METABOLIC RIFT

To conclude, I'd like to review the main points of the late Marx's breakthrough regarding degrowth communism.

Late in life, Marx called for a shift in production to emphasize use-value and reduce both production associated with the creation of surplus value and the work hours that go with such production. He also called for the abolition of the division of labor that robs workers of their creativity. At the same time, he called for the increased democratization of the labor process. Workers would make decisions about production democratically. It doesn't matter that this kind of decision-making process will take more time. Moreover, there will be a sharp increase in the social value assigned to essential work that's needed by society and low in environmental impact.

The result of all this would be the slowdown of the economy. Embedded as we are in a competition-based, capitalistic society, it might be hard to accept deceleration as a positive thing.

Yet it is capitalism, with its demands for unlimited maximization of profits and economic growth, that is fundamentally unable to protect the Earth's environment. Both humanity and nature become objects of exploitation under capitalism. Furthermore, the artificial scarcity created by capitalism renders large parts of humanity destitute.

The society brought about by the decelerated economy of degrowth communism will be, by contrast, one equipped to satisfy people's needs while also expanding the capacity for society to address environmental issues. By democratizing and decelerating production, we can heal the rift in the metabolic interaction between humans and nature.

Of course, this must be a comprehensive project, including the public management of water and power, the expansion of social ownership, the prioritization of essential work, the reorganization of agriculture, and so on.

In the face of this, instances of resistance like the rise of workers' co-ops or the revolt of the caring classes may seem quite small. And indeed, they might be. But there are many similar small instances of resistance to capitalism occurring all over the world. Such isolated incidents have the power to spread until they become a coordinated wave.

In cities all over the world that have been impoverished by global capitalism, a groundswell is rising—everyday people are searching for a new kind of economy that will address their suffering. These movements are occurring in cities in every region of the planet, gathering the power to change politics even at the national level.

These resistance movements are certainly not all calling for degrowth, and neither are they consciously arguing for communism. But movements that hold within them seeds with the potential to grow into degrowth communism are spreading more and more. This is because in the age of climate crisis known as the Anthropocene, any movement facing the Great Divergence of capitalism by envisioning a new society completely different from our current one will inevitably move in this direction.

BUEN VIVIR

This potential can be seen as well in the recent spread of the concept of *buen vivir*. It's an expression that, when translated directly, means "to live well," originating from the Indigenous peoples of Ecuador before being translated into Spanish. When Ecuador's constitution was revised in 2008, this phrase was adopted to describe the state's obligation to guarantee citizens' right to *buen vivir*.

The phrase then spread across South America, and it is now being used by leftists all over the world. Its spread is due in part to the worldwide movement to revise our value systems by rejecting Western ideals of economic growth and learning from Indigenous peoples. Another example of this is Bhutan's Gross National Happiness index (GNH).

In the United States, we can see a similar impulse animating the large-scale resistance movement formed through cooperation between Indigenous and non-Indigenous people to protect sacred water sources during the anti-pipeline protests at Standing Rock. The journalist Naomi Klein, who took part in that struggle, has started calling explicitly for a new system to replace capitalism in her work. I want to draw particular attention to her statement, made during that time, that we must have "the humility to learn from Indigenous teachings about the duties to future generations and the interconnection of all life."[161] Subsequently, she has begun to adopt a position calling for degrowth.

The current climate crisis is providing the impetus for the development of new movements to discard Eurocentrism in favor of learning from the Global South. This is the kind of development the late Marx himself wished for. These seeds of communism have the potential, spurred by the deepening of the climate crisis, to become bolder and more vibrant until they bloom into an environmentalist revolution for the twenty-first century.

In my final chapter, I will draw our attention to a few such seeds of change.

8

The Lever of Climate Justice

INTERPRETING PRAXIS THROUGH THE LENS OF LATE MARX

Seeds of degrowth communism are sprouting all over the world. I want to end this book by looking at some revolutionary initiatives emerging in cities worldwide through the lens provided by Marx's later ideas, as discussed in this book. Using this new lens, we can see what aspects of these movements and actions could be developed further. The insights provided by Marx's thinking later in life allow us to see the world differently than ever before. This, after all, is the true purpose of theory.

Yet theorists can also learn from the hardships and resistance efforts occurring in the real world. Behind Marx's eventual rejection of the progressive view of history and the acceptance of degrowth as a possibility lies his attention to the Global South. It was only when Marx earnestly looked to the Global South that his value system shifted. If Marx had clung to his Eurocentric stance, it would have been impossible for him to reach the insights he arrived at near the end of his life.

Marx's openness to learning from the Global South is even more important for us to emulate now, in the twenty-first century. One major reason is that, as we saw in chapter 1, the climate crisis caused by capitalism is being felt nowhere more intensely, due to displacement and externalization, than in the Global South.

NOT BACK TO NATURE, BUT TOWARD
A NEW RATIONALIZATION

I will say again, just to avoid being misunderstood, that late-in-life Marx was not advocating the rejection of either urban living or technology in favor of returning to communal farming societies. For one thing, such a transition would be impossible at this point. But neither is there any need to idealize such a lifestyle. It's self-evident that communal farming villages had their own problems, while it's also obvious that urban life and technological advancement both have praiseworthy aspects. So there's no reason to reject all forms of rationalization associated with cities and technology out of hand.

But it remains true that the current state of city life possesses a host of problems that need to be addressed. Mutual aid within communities has been thoroughly dismantled, and lifestyles that depend on high levels of resource and energy consumption are unsustainable. We are in a state of over-urbanization.

As a result, cities produce about 70 percent of all carbon dioxide emissions. This is why we must transform urban life to address the climate crisis and restore mutual aid. If we simply ignore cities and retreat into the mountains, the entire Earth will be swallowed by the *déluge* to come anyway and all will be for naught. What's necessary instead is to critique urban space as it has been produced by capital and replace it with new forms of urban rationalization.

Happily, rational ecological municipalist reform movements have been sprouting up in the form of local municipal bodies. Among them, the movement receiving the most attention goes by the name of "Fearless Cities," a global municipalist movement touched off by activists in Barcelona, Spain. I want to look at this effort from a Marxist point of view. In doing so, we will see the truly revolutionary potential of Barcelona for the first time.

THE CLIMATE EMERGENCY DECLARATION
OF THE FEARLESS CITY, BARCELONA

The term "Fearless City" represents an innovative form of local governance that stands in opposition to the neoliberal policies imposed by the state. Unafraid of not only the state but also global industry, it's a municipal body whose mission is to fight on behalf of its residents.

Citizens' groups and political parties from cities all over the world have joined the Fearless Cities network, including Amsterdam and Paris, which have both placed time restrictions on Airbnb stays within their city limits, and Grenoble, which has banned the use of products made by international corporations in meals served by state schools. A single municipal body can do little to fight global capitalism by itself, which is why cities around the world and their citizens are joining up to exchange information and act in solidarity to create a new society for all.

Among these municipalities, the one that stands out for its boldness is Barcelona's city government, which was the first to fly the Fearless City banner. Its revolutionary nature can be seen, for example, in the Climate Emergency Declaration it issued in January 2020.

This declaration is no superficial statement calling for an end to climate change. It's a true manifesto that presents statistical goals for total decarbonization (that is, zero carbon emissions) by 2050 and includes pages of analysis and action plans for reaching these goals. Though Barcelona is a large city, it is not a capital city, and the high level of policy formulation displayed is surprising. Moreover, this document is neither a report produced by municipal workers nor a think tank position paper; rather, it's the result of the collective work done by city residents themselves.

The plan of action included in the declaration is both comprehensive and concrete, consisting of more than 240 items. These include reducing carbon dioxide emissions; greening urban spaces; expanding public transport; restricting the use of automobiles, airplanes, and ships; addressing energy shortfalls; promoting waste reduction and recycling; and many others.

The fearlessness of the Fearless City stance is evident, as the declaration includes many items—such as the prohibition of short-distance airline routes and speed limits for cars of 30 km/hour within city limits—that cannot be implemented without confronting global industry directly. We can see the prioritization of protecting both the environment and citizens' living conditions over promoting economic growth. This demonstrates a clear shift toward prioritizing use-value over commodity value, precisely the transition called for in the previous chapter—one that constitutes the essence of the degrowth communism envisioned by Marx.

We can vividly see a readiness to bring about a degrowth society in the declaration's action item "Change of Economic Model":

> The current economic model is based on continuous growth and a never-ending race for profits, with an ever-increasing consumption of natural resources. The same economic system that is putting our planet's ecological balance in danger has significantly increased its inequalities. Without a doubt, the global ecological crisis and the climate crisis in particular are largely due to excessive consumption on the part of the rich countries and, above all, the wealthiest social groups.[162]

Capitalism is being criticized here in harsh and explicit terms for giving rise to the "never-ending race for profits" and excessive consumption of resources directly responsible for climate change. This radical statement was put forth by common citizens, gained support with the populace, and eventually accumulated enough power to spur the city government into action. This chain of events holds within it hope for the future.

FROM SOCIAL MOVEMENT TO LOCAL POLITICAL PARTY

Of course, Barcelona's groundbreaking declaration didn't appear overnight. A strong, tenacious citizen movement lasting more than ten years led to this point.

As is well known, Spain was one of the countries hit hardest by the EU's financial crisis touched off by the 2008 collapse of Lehman Brothers. At the time, unemployment hit 15 percent, and poverty became rampant at the same time that public services and social security were being cut mercilessly due to austerity measures imposed by the EU.

Adding insult to injury, the lives of average citizens in Barcelona grew more difficult due to the overdevelopment of the tourism industry. Landlords were methodically replacing rentals catering to locals with short-term "lodgings" catering to tourists. Rents skyrocketed, leading many residents to lose their homes. The cost of living also increased precipitously. Barcelona became a city where the contradictions of neoliberal globalization reared their heads most destructively.

An anti-austerity movement led by young people impoverished by this situation arose in 2011. Called the 15-M Movement, it initially took the form of occupations of the city's public plazas. This movement took many other forms as it continued, until it resulted in the formation of a locally grounded municipal political party called Barcelona en Comú—"Barcelona in Common."

Candidates from this newly formed party ran in the 2015 local elections, leading one of its central figures, Ada Colau, to be elected mayor. A social activist engaged with anti-poverty movements, she was most well known for advocating for housing rights.

As a new mayor who maintained her connection to the movement that put her in power, she set up a system to bring grassroots voices into city government. She carefully listened to neighborhood associations and those who worked in the realm of the commons—waterworks, energy, etc. She opened the city hall to common citizens, making city council meetings function as platforms for citizens' voices to be heard. We can see here a superlative example of politics arising directly from connections to social movements.

The process of drafting the declaration was conducted in the same manner. It was based on analysis conducted by a Climate Emergency Committee made up of more than three hundred citizens drawn from more than two hundred citizens' groups and associations. Barcelona Energia, a public enterprise focusing on renewable energy, and other people engaged in areas like public housing also joined workshops leading up to the writing of the declaration.

In other words, the declaration was cowritten by citizens, workers, and specialists from all areas related to social production. It was a project conducted jointly by citizens from all walks of life in the city. If it hadn't been conducted this way, it surely wouldn't have been able to achieve the level of concreteness that it did as a reform package. As Marx would say, true social transformation arises from the realm of production.

COUNTERING CLIMATE CHANGE CREATES HORIZONTAL SOLIDARITY

Barcelona, of course, was already home to social movements and projects of all sorts, addressing areas like water, power, housing, and the like. But it took climate change to knit these various single-issue movements (for example, the movement to make water publicly owned and operated) into a mutually beneficial network. Applying the perspective of climate change to single-issue reform movements allows for the creation of horizontal solidarity that transcends the particularities of specific issues.

For example, rising electricity bills have hit the poorest residents of the city hard. If the energy provider were replaced by a publicly owned renewable energy company that aimed to deliver local production for local consumption, it would stimulate the local economy, allowing any profits gained to be reinvested into the community. This, of course, would help fight climate change and would also help address poverty. As another example, constructing public housing fitted with solar panels would help the environment while also helping to create housing security for residents and combating the forces of capital arriving in the form of gentrification. The creation of an economy focused on local production for local consumption helps create new employment opportunities for local residents while also easing the unemployment problem among the youth.

Incorporating efforts to address climate change into various single-issue movements allows for connections to arise between them, driving them to push for more systematic change, including economic, social, and cultural transformations.

On top of that, what we see here in these examples is the conscious replacement of the artificial scarcity of capitalism with the radical abundance of the commons.

PARTICIPATORY SOCIETY REALIZED THROUGH COOPERATIVES

One secret to the continued success of the Barcelona movement at the level of both the policy and methodology is the long tradition of workers' cooperatives there—the very same workers' co-ops that Marx called examples of "possible" communism.

Spain has always been a hotbed of cooperative associations, and Barcelona is famous for its social solidarity economy, which includes not only workers' cooperatives but also consumer cooperatives, mutual aid societies, organic produce collectives, and so on. The social solidarity economy, in fact, employs about 53,000 people, or 8 percent of the people employed in the city, producing 7 percent of the city's total gross production.[163]

The range of activities undertaken by the city's workers' co-ops is expansive, encompassing industries of all kinds, including manufacturing, agriculture, education, waste disposal, housing, and more. By offering vocational training for youth as well as support for the unemployed and local residents, they've also helped clear the path toward local citizen-led urban renewal projects to combat gentrification and over-tourism.

Connecting municipal governments and cooperatives produces positive results for both. It leads municipal governments to prioritize local, justice-oriented projects when allocating public support and funding and allows cooperatives to flourish as they gain more sources of funding. It's a transition from outsourcing to insourcing.

For their part, cooperatives find that their voices finally reach decision makers at city hall, stimulating both political engagement and social movements. Shifting the emphasis away from short-term profit making and toward mutual aid and the autonomous, active participation of cooperative members in politics paves the way for participatory socialism to cross over from the site of production to society at large. A dynamism arises between citizens and local governance that was previously absent, spurring both to higher levels of performance.

This is the first step toward the transition to a form of sustainable participatory socialism based on mutual aid, not an economic model based on exploitation and plunder. This is what Marx referred to as "free association."

TOWARD AN ECONOMIC MODEL THAT BRINGS ABOUT CLIMATE JUSTICE

Now I would like to touch on what I consider to be the most ground-breaking part of Barcelona's bold Climate Emergency Declaration. The document makes a point of emphasizing the following: that large urban centers in the developed world must accept the magnitude of their responsibility in causing climate change as a crucial first step in the realization of climate justice.

As we have seen, the wealthy classes in developed countries bear the most responsibility for bringing about climate change, yet the communities most affected are those in the Global South, where the use of fossil fuels has been comparatively modest, as well as future generations. The general term for this group is MAPA—"Most Affected People and Areas." Climate justice is the call to stop climate change while alleviating the injustice the MAPA are subjected to.

The Barcelona Declaration states that the transition to an economic system that can bring about climate justice must include the voices of those most severely affected by environmental damage:

specifically women from the Global South. The document states: "Women are disproportionately more vulnerable to the impacts of the climate crisis. In fact, 80 percent of people displaced by the climate crisis are women, yet they are the main caregivers. If we are to tackle the climate emergency, we have to transform an economic model that is unsustainable and unfair."

Moreover, the document continues, major cities in the developed world have the responsibility to lead the transition to a society that leaves no one behind, that promotes both "shared care work" and the "fraternal relations" between human beings, "other living beings," and nature. The burden of this transition must be borne not by the most vulnerable but by "the most privileged."[164] This is nothing less than the revolt of the caring class.

MUNICIPALISM TRANSCENDING NATIONAL BORDERS

What's most important to note here is that the movement in Barcelona is not limited to a single city in a developed country but rather directs attention to the Global South as well. This has led the movement to foster international solidarity in the fight against the dominion of capital.

Indeed, after arising in Barcelona, the Fearless Cities network now includes seventy-seven locations, including ones in Africa, South America, and Asia. The fearlessness of the Fearless Cities movement is enabled not only by mutual aid within the cities but also by cooperation between cities.

For example, knowledge is being shared between cities about how to reverse the privatization of public services like waterworks that was rampant during the period of neoliberal policymaking. The private businesses running waterworks are frequently large transnational corporations, and negotiating with them can be contentious, sometimes involving litigation, so sharing know-how through Fearless City networks can be a real lifesaver in these efforts.

The spirit of networking and solidarity across national borders between reformist municipal bodies is referred to as "municipalism."[165] In contrast to previous forms of municipal governance that tended to be closed and provincial, municipalism strives toward city governance that is open and international.

LEARNING FROM THE GLOBAL SOUTH

This is not to say that the municipalist movement has always been perfect. As a project launched from Europe, municipalism has faced criticism from the Global South. Isn't this just another movement centering white people from the Global North?

The fact is, both communal management and participatory socialism independent of the state are efforts that developed first and most effectively in the Global South. The most famous example is the Zapatista movement led by the Indigenous peoples of Chiapas, Mexico. The movement arose in 1994, when the North American Free Trade Agreement (NAFTA) was passed. The Zapatistas clearly said no to neoliberalism and global capitalism well before the rise of municipalism in Europe.

To take another example, this one of a movement striving for solidarity across national borders, the international farmers' cooperative Via Campesina (Spanish for "The Peasants' Way") emerged around the same time as the Zapatista movement. Arising in 1993 in response to the rapid deregulation of agricultural produce markets, it consists predominately of indigenous women-led organizations located in Central and South America. These are undeniably voices from the Global South.

The call to take back the business of agriculture and manage it autonomously is, naturally, one related directly to survival. It's a demand for what's known as food sovereignty.

The traditional farming practices and agro-ecological reforms advocated by the small and midsize farms that make up Via Campesina are also relatively light in terms of their burden on the environment. The 1990s, when this cooperative began, was a time of rapid rises in carbon dioxide emissions following the end of the Cold War. The flip side of the rise of global capitalism at that time was the concomitant rise of revolutionary resistance movements in the Global South like the Zapatistas and Via Campesina.

It's therefore natural for those in the Global South to point out that it was the Global North that slept on the destruction of the environment that took place at that time. It would make more sense for organizations in the Global North to recognize the pioneering efforts by those in the Global South and learn from them rather than the other way around.[166] How many people in the Global North are even aware of the Via Campesina movement, despite it including more than 200 million farmers from all over the world?

THE INEFFECTIVENESS OF ENLIGHTENMENT

This book began with an analysis of the Imperial Mode of Living and ecological imperialism. The luxurious, rich lifestyles enjoyed by those in the developed world are only possible due to the extraction of resources from and the externalization of environmental burdens onto the Global South.

The externalization society that imposes its environmental impact onto the Global South—that is, our society, the society of the Global North—closes its eyes to this injustice, absorbed in its capitalist dreams and uninterested in discovering what's really happening in the rest of the world.

This is why we must address ecological imperialism and the Imperial Mode of Life if we want to bring about a truly just and sustainable new society. We'll never solve the problem by changing the consumption patterns of those living in the Global North one by one. An enormous change must occur on a global scale.

But it's clearly also insufficient to take on the cosmopolitan pose of the "world citizen" and call for "enlightenment" in the face of the exploitation and plunder of the Global South. In this context, declarations of abstract ideals to address horrifying realities can't help but ring hollow.

Instead, we must turn our attention to the real resistance to exploitation that's already occurring. Most crucial in this effort is examining the concrete reasons why efforts to construct economies based on international solidarity arose in the first place.

This is, as we've seen, the exact sort of project Marx embarked upon at the end of his life. He noticed that it was in areas at the margins of capitalism—the areas we now call the Global South—that its cruelty showed its face most clearly.

This is why Marx mined the possibilities of anti-capitalist examples like the communal farming villages of Russia and the anti-colonialist movements of India. The culmination of this intellectual project was, as we saw in chapter 4, the proposal of degrowth communism.

In the same way, the municipalism of local governments aiming to bring about a just, sustainable society should respond to critique and actively learn from resistance movements in the Global South. The core movements there are those calling for climate justice and food sovereignty.

RESTORING FOOD SOVEREIGNTY

Let us first examine the issue of food sovereignty.

It's obvious that food is necessary for life and should therefore be considered a form of commons. Capitalistic agribusiness in the Global South, though, is predicated on exporting crops to the Global North. This is why many countries, despite having thriving agricultural sectors, are producing crops purely for export and therefore have large populations of people who suffer from hunger.

This issue arises from the way crops that can be exported at high prices to decorate the tables of the developed world are prioritized over the cheaper produce that those actually cultivating the fields need to live. Furthermore, the monopolization of information and the right to use fertilizer, pesticides, and even seeds by multinational corporations via the patent system render the financial burdens on farming families crushingly heavy.

In this way, the contradiction within capitalism by which the production of use-value is denigrated in favor of the production of value emerges in the cruellest of forms in the Global South.

To take a specific example, the legacy of the twisted system of apartheid, which stemmed in part from the Dutch and English colonization of South Africa, has resulted in less than 20 percent of the farms in the country—vast agricultural businesses that are mostly white-owned—producing 80 percent of its total agricultural goods. Despite being one of the biggest exporters of food in Africa, South Africa's hunger rate exceeded 26 percent in 2020.[167] Under apartheid, small-scale farmers of color were relegated to less fertile land with poor access to water, rendering it difficult to produce enough to support even themselves. This is happening despite South Africa being added to the BRICS group of fastest-developing countries in the world in 2010.

In 2015, a citizen-led movement was launched to address the situation called the South African Food Sovereignty Campaign (SAFSC).[168] Participants include owners of small-scale agribusinesses and farmworkers as well as leaders of local NGOs and social movements. The campaign created a platform for promoting cooperative farming at a grassroots level. This is a revolt against the top-down agribusiness led by the national government, which had so clearly failed to enrich the lives of its own citizens.

The main issue addressed by the campaign is the lack of knowledge about sustainable farming practices among poor farmers. As it was, these farmers found themselves having to borrow money to buy chemical fertilizer and pesticides, making it easy for them to fall prey to large-scale agribusiness.

Under the SAFSC model, farmers create cooperatives on their own terms. Local NGOs lend out farming tools and educate farmers on organic farming practices. In other words, they are carefully conducting the kind of vocational training Marx saw as necessary to return the skills and technology monopolized by capitalists to the producers.

The campaign also intends to allow farming practices that will enable farmers to cultivate crops without depending on genetically modified seeds and chemical fertilizer. In other words, the SAFSC is at its core an effort to restore the commons.

FROM THE GLOBAL SOUTH TO THE WORLD

Movements like Via Campesina and SAFSC are well aware that fighting for food sovereignty alone is insufficient. There's a larger issue at hand, namely the subject of this very book: climate change.

South Africa's agriculture is in fact being threatened by climate change right now. Cape Town has been chronically plagued by serious water shortages. It's predicted that from this point on, the risk of drought is extremely high. Rises in food prices due to chronic drought will make life extremely difficult for the average citizen.

This is why it's insufficient for these movements to aim to stabilize farming and make it sustainable. Such efforts will be rendered moot in a world where farming itself is impossible. For this reason, movements for food sovereignty are inextricably linked to movements for climate justice. This is why local movements are connecting with each other all over the world.

CHALLENGING THE IMPERIAL MODE OF LIVING

Sasol is a chemical and energy company headquartered in Johannesburg that handles coal, oil, and natural gas. This company is responsible for approximately 67 million tons of carbon dioxide emissions every year[169]—more than the total yearly emissions of Portugal. The air pollution caused by Sasol is also, naturally, quite extensive.[170]

Why are its emissions so high? One reason is that it's engaged in the creation of synthetic oil refined from coal as a replacement

for regular oil. During the apartheid era, South Africa was subject to an oil embargo. A nationally owned company at the time,
Sasol responded to the situation by using the Fischer-Tropsch
process developed in Germany during the Nazi era to synthesize
oil instead.

Today, South Africa can import crude oil again, but the company
has continued using this process to create synthetic oil anyway,
drawing the attention of the world once more. Even as oil sources
are drying up, coal is still plentiful. The technology Sasol has developed for synthesizing oil from coal has naturally made it a darling
in the energy world. Yet the use of synthetic fuel refined out of coal
results in almost twice the greenhouse gas emissions of using regular oil.[171] From the point of view of the climate crisis, this technology
passes the buck to a lethal degree.

Environmental activists in South Africa are naturally calling for
Sasol to stop this incredibly burdensome operation.[172] What's most
interesting, though, is the method employed by this movement.
Vishwas Satgar, a central figure in the SAFSC, emphasizes that the
movement is not primarily a South African one but rather one
demanding solidarity among movements internationally under
the slogan "We Can't Breathe!"

What Satgar and his allies are focusing on is Sasol's investment
in the creation of an oil and chemicals plant in Lake Charles, Loui-
siana (called the Lake Charles Chemical Project, or LCCP). This
project will naturally result in large amounts of emissions in the
United States as well.

For this reason, Americans concerned about climate change
should see the demand for Sasol to cease its operations as something
that affects them directly,[173] prompting solidarity from social
movements like the Sunrise Movement, Fridays for Future, and
Black Lives Matter.

Indeed, this is a call for solidarity that goes beyond the international movement to reduce carbon emissions. It's a call from the
Global South directed at the developed world to reflect on the history
of imperialism linking the European colonization of South Africa
and the creation of apartheid to the oil industry in the United States

and to take this opportunity to break away from the legacy of capitalistic burden that comes with it. In other words, it's a call for global solidarity in challenging the Imperial Mode of Living.

We can see this in the adaptation of the "We Can't Breathe!" slogan from the Black Lives Matter slogan "I Can't Breathe!," the latter being the last words spoken by Eric Garner, who died when NYPD officer Daniel Pantaleo put him in a choke hold in 2014.

This environmental movement in South Africa proclaims that similar forms of violence occur every day all over the world, and that includes violence inflicted by the air pollution produced by the petrochemical industry. Furthermore, it connects the climate change problem to the legacies of imperialism and racism that found their purest expression in the slave trade, expanding activism in these areas into the context of climate justice.

Human rights, the climate crisis, gender—all these issues are connected through capitalism.

South Africa is not the only place where this call is being made. Various movements around the world are taking it up as well. It's only that we in the developed world aren't noticing it or choose to ignore it. But if we don't answer the call, climate justice will never be realized.

Late in life, Marx criticized England's colonial rule in Ireland and urged British workers to act in solidarity with the oppressed Irish people. As Marx put it, "the lever must be applied in Ireland"—that is, British workers will never be liberated if the Irish people aren't freed from their oppression.[174]

In exactly the same way, the revolutionary "lever" of today must be applied first in the Global South. But is the necessary solidarity possible?

THE LEVER OF CLIMATE JUSTICE

In fact, the Climate Emergency Declaration of Barcelona examined earlier is a prime example of just such an effort to answer the Global South's call for justice. What is even more interesting is that the actions taken to respond to this call are essentially the steps one would take to transition to degrowth communism.

As I pointed out before, the Barcelona Declaration clearly articulates the injustice of socially vulnerable people in Global South countries bearing the brunt of climate change when the carbon dioxide that brought about this disaster was emitted, for the most part, by the Global North. Furthermore, it states that major cities in developed countries bear the most responsibility for addressing the problem, but in a way that fights for climate justice not just for those in their own country but for all, so that "no one's left behind."

Just as Marx learned about degrowth through his study of non-Western and precapitalist societies, Barcelona seems to have learned about climate justice from the Global South. This is linked to the groundbreaking nature of the declaration. In essence, Barcelona is using climate justice as the lever for revolution.

Why is climate justice so important to this discussion? I would like to remind us here of the arguments I made in chapter 2 and chapter 5 of this book. Thomas Friedman, Jeremy Rifkin, and even Aaron Bastani all call for the transition to a sustainable economy. But in my view, their prioritization of economic growth guarantees that the results of their recommendations would simply be the intensification of the plunder of the periphery.

What I believe all these thinkers lack most fundamentally is a perspective on the Global South—or rather, the willingness to learn from the Global South.

Global North countries have always been able to balance environmental problems and economic growth and may seem able to continue doing so indefinitely. But beneath this seeming balance is the displacement of various problems onto the Global South, rendering them invisible. This is why even if countries in the Global South try to emulate the Global North in their quest to balance environmental issues with economic growth, it will never work. There's nowhere left to displace growth's costs. Our current climate crisis demonstrates that we've already reached the outermost limits of externalization society.

We could always turn our backs on the crisis and, in the manner of Friedman and Bastani, proclaim that decoupling is possible and that capitalism-driven dematerialization will solve the climate issue for us. But we can also take the idea of climate justice seriously, turn our attention to the Global South, and learn from the various efforts taking place there. Only then will we be able to grasp what might truly need to happen to realize a future that's both sustainable and just.

BARCELONA AIMING FOR DEGROWTH

Barcelona is, of course, calling for large-scale reforms of infrastructure like the adoption of solar power and the introduction of electric buses. Increased public spending on anti-austerity measures is also necessary. But from the point of view of climate justice, these large-scale reforms must never come at the expense of the natural environment or the people of the Global South. To prevent such damage from being inflicted, capitalistic economic growth must be brought to an end once and for all.

This is why the Barcelona Declaration, in lieu of celebrating "green growth," explicitly criticizes the "continuous growth and a never-ending race for profits" that characterizes the present system.

In other words, the difference between the Green New Deals trumpeted by Thomas Friedman and his ilk and the Barcelona Declaration is, essentially, the difference between growth- and degrowth-based worldviews. Precisely because it incorporates an openness to learning from the Global South, the latter's vision of a future sustainable society is fundamentally different from the former's.

Isn't Barcelona's way of doing things quite similar to the way envisioned by Marx? They are learning from the Global South as they attempt to explore possibilities for new forms of international solidarity. Doing so will inevitably lead to the discarding of the productivism that insists on economic growth in favor of envisioning a society that prioritizes use-value above all else.

THE PROBLEM WITH THE CURRENT LEFT

Compared with Barcelona's efforts to bring about climate justice, conventional Marxists' theoretical fixation on growth stands out in stark relief. Socialism attempted to do away with exploitation. But this ended up being merely an effort to bring about a society in which a nation's proletarian class could enjoy the material abundance realized by capitalism.

A future society realized in such a way would, outside of the absence of capitalists, differ little from our present one. And indeed, in the case of the Soviet Union, the management of nationalized industry by bureaucrats and officials eventually resulted in something that should probably be called "nationalized capitalism."

Marxism conceived in this way can never offer a truly radical solution to the crisis of the Anthropocene. The result of this is the continued neglect of Marxist thought even as the contradictions of capitalism deepen so obviously and so destructively.

The current left is preoccupied with fighting neoliberalism, which, to be fair, represents an attempt to intensify the exploitation of workers even more. The policies promoted by neoliberalism, including instituting austerity measures, destroying the social safety net, lowering wages by increasing freelance and temporary work, dismantling public utilities through privatization, and so on are all things that have made the quality of our lives worse.

But is it sufficient to respond by demanding that states increase public funding and redistribute wealth under the banner of anti-austerity and returning wealth to the hands of workers? Of course, if we manage to ride out the present long-term stagnation and get the world economy running again, such a future would be an improvement on our present misery.

Stopping at calling for the end of austerity, though, does nothing to halt the plunder of nature. We'll never survive the crisis of the Anthropocene by stimulating the economy.

TOWARD RADICAL ABUNDANCE

There's another problem with current leftist thought. People calling for the end of austerity seem to believe it's neoliberalism's austerity measures that lie at the root of the present scarcity. If that were true, then abundance could be achieved by stimulating production through public investment, which would increase accumulation and grow the economy. But this is a capitalism-friendly mode of thinking. In other words, what appears at first to be a radical form of leftism reveals itself to be a fundamentally conservative way of thinking, one that seeks only to prop up the status quo.

This level of reform is inadequate. The cause of scarcity isn't neoliberalism, it's capitalism. We need to move beyond policy change and toward changing the social system as a whole if we want our efforts to be adequate in the age of climate crisis. The true revolutionary plan proposed by Marx late in life was the complete escape from capitalism and the restoration of radical abundance, which can only be brought about through degrowth.

REJECTING THE POLITICS OF BUYING TIME

This is why I argue in this book for the possibility of revolutionizing the sites of production while taking care to restore the commons. I've also critiqued the politicalism that relies on top-down policies, laws, and systemic change as the path to social revolution. As I've said, politics aren't independent of the economy—it's subordinate to it.

I want to emphasize here that the main problem with top-down politicalism is the extreme narrowness of the political choices possible within the present scheme. As we've seen, Green New Deals promoting "green growth," dream technologies like geoengineering, and economic policy theorizations like Modern Monetary Theory all propose what seem like unconventional, revolutionary large-scale transformations to address the coming climate crisis, but in the end, all of them still support the root cause of that crisis—capitalism. This is a severe contradiction.

All these kinds of political measures can do is buy a little time to actually solve the problem. But in the current ecological situation, such efforts to buy time are themselves fatal. The most dangerous thing would be for people to be lulled into complacency by policies that only seem like solutions and stop thinking seriously about the crisis at hand. This is why the SDGs proposed by the UN need to be criticized as well. It's time to dispense with half measures and move down the path to the social ownership of the oil industry, major banking, and the digital infrastructure currently monopolized by the GAFA constellation. In short, there needs to be a revolutionary transition to communism.

But what good does it really do to criticize politicians in the Global North? Even if they propose policies to combat climate change, they can hardly expect votes from the Global South or future generations. Politicians are necessarily creatures who cannot think about problems outside their relevance to the next election. Furthermore, their decision making is hindered by donations and lobbying by major industry. Therefore, if we want to truly face the climate crisis, we must reform democracy itself.

REFORMING THE TRINITY: ECONOMICS, POLITICS, AND THE ENVIRONMENT

The reform of democracy is more important now than ever before. This is because the power of the state will play an indispensable role in any effective response to climate change.

I have argued in this book that the foundation of communism is the equal communal management of the means of production as a form of commons—that is, as something distinct from private ownership or ownership by the state. This does not, however, mean that I reject the power of the state entirely. Indeed, it would be foolish to reject the state as a means of getting things done, such as the creation of infrastructure or the transformation of production. Anarchism, which does reject the state, cannot effectively combat climate change. But depending too much on state power may easily lead to a descent into climate Maoism. This leaves communism as the only real choice left.

At this juncture, in order to avoid a top-down, totalitarian situation controlled by politicians and technocrats, we must systematize a process whereby the state's nature reflects the views of its people and fosters citizens' participation and agency.

By expanding the realm of the commons even as we acknowledge the power of the state, we can open access to democracy beyond the walls of the legislature and into the realm of production. As we saw in chapter 6, "private citizen-ization," along with cooperatives and social ownership, can serve as examples of what this might look like.

At the same time, democracy itself must undergo a major shift. As we've seen, municipalism is just such an effort, one that's taking place at the local level. At the state level, citizens' assemblies like those discussed in chapter 5 can serve as a possible model.

As restoring production as a form of commons, municipalism, and citizens' assemblies as forms of democracy that truly allow citizens to exercise their agency in planning the way forward expand their reach, even more fundamental debates will begin to occur as to what sort of society we all want to live in going forward. In other words, society will evolve so that everything—from the meaning of work to the meaning of life to the meaning of liberty and equality— can be debated from the ground up, openly and freely.

Questioning the meaning of everything from the ground up will inevitably overturn all that is taken for granted now as "common sense." This is the moment when the truly political, that which transcends all existing frameworks, can finally emerge.

This is the revolutionary trinity: overcoming capitalism, reforming democracy, and decarbonizing society. The expansion of synergy among the realms of economics, politics, and the environment is the only thing that can bring about a truly fundamental transformation in our social system.

THE GREAT LEAP TOWARD A
SUSTAINABLE, JUST SOCIETY

The basis of such a project must be mutual aid and trust. Without these, the only solutions possible are undemocratic and top-down.

We are living in an era where mutual aid and trust in others has been thoroughly dismantled by the forces of neoliberalism. The only way to rebuild these trust relationships is through face-to-face community building and local municipal politics, at least at the start.

There are surely those who say that such humble actions will never bring about change in time. But communities, regional associations, and social movements whose reach seems restricted to the local are finding ways to link up with comrades all around the world in solidarity, and it is here that hope for the future resides. We are already seeing how various local movements are beginning to construct networks with other movements around the world to fight global capitalism. As Via Campesina puts it, "Globalize the Struggle, Globalize Hope!"[175]

Such international solidarity will provide people with the experience of confronting capital and thus give them more power, inevitably changing their worldviews and values as well. As the power of people's imagination expands, we will become capable of acting and thinking in ways not yet conceivable in the present.

As community and social movements gain strength, politicians will become afraid not to instigate greater and greater changes. Barcelona's city government and France's citizens' assemblies are cases in point.

This will spur politics and social movements into greater levels of interaction. It will also allow the bottom-up organization of the social movement to join forces with the top-down organization of party politics, maximizing the power of both. A completely different form of democracy from that envisioned by politicalism will emerge.

Indeed, in Spain, the movement in Barcelona has created waves that have reached all the way to the national level. The politician Alberto Garzón, who had served as the general coordinator of the

United Left (UI) party, went on to serve as the minister of consumer affairs in the second government of Pedro Sánchez. In May 2022, Garzón presented a paper titled "Limits to Growth: Eco-socialism or Barbarism?" In it, Garzón clearly rejects green economic growth and asserts that democracy has no future without a new form of socialism. He further asserts the necessity of degrowth. In short, he articulates much the same position as this book.

More concretely, Garzón has called upon the nation to decrease its consumption of factory-farmed meat and reduce airplane use. It was inconceivable only a few years ago that a politician would make these sorts of demands, but the tide is shifting in the face of Generation Z's insistent call for action.

This trend will only grow stronger. As it does, so will the power to bring about a clean break with the false dream of unlimited growth and make a great leap toward the creation of a sustainable, just society for all. A door that was once closed will open.

Where will this great leap take us? To a future founded on mutual aid and self-governance—to a future of degrowth communism.

CONCLUSION
How to Prevent the End of History

Degrowth through Marx? Are you crazy?

I began writing this book well aware that I'd face this sort of disbelieving response from all quarters.

The left states that Marx never called for degrowth or anything like it. The right just scoffs, wondering if I'm aiming to repeat the failures of the Soviet Union. On top of this, reflexive rejection of the very concept of degrowth runs deep even in liberal circles.

Yet I couldn't go on without writing this book. I'm convinced that the best way to survive the crisis of the Anthropocene can be found in the insights Marx had near the end of his life, which can be summed up as degrowth communism. I arrived at this conviction while analyzing the latest research into Marx's thought along with the connections between capitalism and the climate crisis.

If you've been kind enough to read all the way to the end of this book, I hope you're now able to see that the only hope humanity has left for surviving the climate crisis and bringing about a sustainable, just society is degrowth communism.

As we saw in some detail in the first half of this book, climate change cannot be stopped by SDGs, Green New Deals, or bioengineering. "Green Keynesianism" pursuing "green growth" will only result in the further entrenchment of the Imperial Mode of Living and ecological imperialism. All that will happen then is the further spread of inequality, accompanied by the worsening of the global environmental crisis.

We cannot solve a problem triggered by capitalism while still preserving capitalism, as there is no other root cause. We must make a thorough break with capitalism to find a solution to climate change.

Furthermore, it's precisely capitalism, with its drive for unlimited profits reaped from artificial scarcity, that's making our lives miserable. By restoring the commons dismantled by capitalism, degrowth communism has the potential to make a richer, more human way of living possible again.

If we insist on extending capitalism's life, we're dooming ourselves to a descent into barbarism brought about by the chaos of the climate crisis. Right after the Cold War ended, Francis Fukuyama famously declared the "end of history" while postmodernism proclaimed the waning of "grand narratives." What has become clear in the course of the thirty years since is that what really awaits us after spending this time laughing cynically in the face of the threat posed by capitalism is a completely unforeseen "end of history"—an end to civilization as we know it. This is why it's imperative that we join in solidarity to put the brakes on capital and forge a future of degrowth communism together.

The fact remains that we've become completely used to our lifestyles steeped in capitalism. Even those who largely agree with the broad outlines of the arguments and facts presented in this book will likely still conduct their lives as usual, unable to conceive of what they might do in the face of a demand as enormous as changing an entire social system.

It's true that combating capitalism and the superrich 1 percent who control so much will take more than buying a few eco-bags and reusable water bottles. The struggle will surely be difficult. You might even think that getting 99 percent of the population to take part in a plan that has no guarantee of working is absolutely impossible.

But it would behoove us to remember the figure "3.5 percent." This is the number that Harvard political scientist Erica Chenoweth came up with in the course of her research into protest strategies as the percentage of a population that must rise up sincerely and nonviolently to bring about a major change to society.[176]

Examples of nonviolent civil disobedience by 3.5 percent of a population touching off a major social revolution include the 1984 "People Power Revolution" that took down the Marcos regime in the Philippines and the 2003 "Revolution of Roses" in Georgia that culminated in the resignation of then-president Eduard Shevardnadze.

The Occupy Wall Street movement and the sit-ins in Barcelona also began as protests involving relatively few people. Greta Thunberg's school strike famously started as a one-person protest. Even at its height, the number of committed participants in the Occupy Wall Street protests and sit-ins that gave birth to the slogan "1 percent vs. 99 percent" surely only numbered a few thousand.

Yet these resistance movements have had major impacts on society. Demonstrations have numbered in the several thousands to millions. Videos of these movements have been shared millions and even billions of times on social media. If it were an election, this would translate into billions of votes. This is the road to revolution.

It seems entirely within the realm of possibility that enough people sincerely concerned with climate change and passionately committed to fighting it could gather together to form a constituency of 3.5 percent. There would be even more if we factor in those angered by the environmental destruction and inequality brought about by capitalism and who possess the imagination to want to fight on behalf of the Global South and future generations. These are people whose convictions could lead them to actions that would even make up for those who, for whatever reason, are currently unable to swing into action themselves.

A workers' co-op, a school strike, an organic farm—it doesn't matter the form it takes. You might run for office to become a part of the municipal government. You might act as part of an environmental NGO. You might start a citizen-run electric company with your neighbors. It would be a major step to demand that the enterprise that employs you put in place strict environmental policies. Bringing about the democratization of production and the shortening of work hours, for example, must include the participation of labor unions.

Signature-collection actions should be started that lead to more declarations of climate emergency; movements must be developed to demand that the richest elites pay their fair share. So doing, mutual aid networks will arise and be forged into something truly mighty.

There are so many things that can and must be done right now. The vast scale of systemic change is no excuse for doing nothing. The participation of every individual is decisive in forming the necessary 3.5 percent.

Our indifference up till now has allowed the 1 percent of superrich elites to change the rules as they see fit and organize society to their benefit according to their worldview.

But now is the time to say no! We must shed our postures of cynical indifference and show them the power of the 99 percent. The key will be the actions undertaken now, at this moment, by the first 3.5 percent. These actions will combine to become a huge groundswell that will rein in the power of capital, reform democracy, and decarbonize society.

At the beginning of this book, I stated that the Anthropocene was an era when the man-made products of capitalism, its burdens and contradictions, had overrun the entirety of the Earth. As it's capitalism that's destroying the planet, perhaps a better name for this era would be the Capitalocene.[177]

If that's the case, then it will only be once people band together in solidarity to rein in capital and protect the planet, our only home, that this new era will deserve the name Anthropocene. This book is meant to be a version of *Capital* for this new era, a thorough critique of capital illuminating the path to the bright future to come.

Of course, though, this bright future can only come about if the readers of this book decide to add themselves to the 3.5 percent and help make the changes necessary to realize it.

Notes

Chapter 1

1 Jason Hickel, "The Nobel Prize for Climate Catastrophe," *Foreign Policy*: foreignpolicy.com/2018/12/06/the-nobelprize-for-climate-catastrophe/ (last access on 5/15/2020).

2 William D. Nordhaus, "To Slow or Not to Slow: The Economics of the Greenhouse Effect," *The Economic Journal*, 101, no. 407 (1991): 920–37.

3 William D. Nordhaus, *The Climate Casino: Risk, Uncertainty, and Economics for a Warming World* (New Haven: Yale University Press, 2015), 76. As the argument in this book demonstrates, Nordhaus later became stricter in his recommendations for how to combat global warming, but still with an aim to keep the rise in temperatures between 35.6°F and 37.4°F, rather than between the more accepted 34.7°F and 35.6°F range. Not only that, he even declares the 35.6°F goal to be "not very scientific."

4 Nina Chestney, "Climate Policies Put World on Track for 3.3C Warming: Study," *Reuters*: reuters.com/article/us-climate-changeaccord-warming /climate-policies-put-world-on-track-for-3-3cwarming-study -idUSKBN1OA0Z2 (last access on 5/1/2020).

5 Will Steffen et al., "The Trajectory of the Anthropocene: The Great Acceleration," *The Anthropocene Review*, 2, no. 1, 2015.

6 thechemicalengineer.com/news/dam-collapse-tragedy-could-have-been -prevented/.

7 scielo.br/j/vb/a/dbhfR3vBCpkvW7WRZrjHgVJ/?lang=en.

8 Ulrich Brand & Markus Wissen, *Imperiale Lebensweise: Zur Ausbeutung von Mensch und Natur im Globalen Kapitalismus* (Munich: oekom, 2017), 64–5.

9 The film *The True Cost* (dir. Andrew Morgan, 2015) provides a good overview of the problems associated with this issue.

10 Stephan Lessenich, *Neben uns die Sintflut: Wie wir auf Kosten anderer leben* (Munich: Piper, 2018), 166.

11 Wolfgang Streeck, *How Will Capitalism End?: Essays on a Failing System* (London: Verso, 2016).

12 Wallerstein's thought has obviously influenced that of myriad subsequent thinkers analyzing the plundering of resources from the natural world. One such study is Stephen Bunker's classic article about issues related to the Amazonian rainforest in Brazil: Stephen G. Bunker, "Modes of Extraction, Unequal Exchange, and the Progressive Underdevelopment of an Extreme Periphery: The Brazilian Amazon, 1600–1980," *American Journal of Sociology*, 89, no. 5 (1984): 1017–64. This subsequently led to an approach focusing on "ecologically unequal exchange." Representative works in this line of thought include: Alf Hornborg, "Towards an Ecological Theory of Unequal Exchange: Articulating World-System Theory and Ecological Economics," *Ecological Economics*, 25, no. 1 (1998): 127–36; and Andrew K. Jorgenson and James Rice, "Structural Dynamics of International Trade and Material Consumption: A Cross-National Study of the Ecological Footprints of Less-Developed Countries," *Journal of World-Systems Research*, 11, no. 1 (2005): 57–77.

13 Markus Gabriel, Michael Hardt, Paul Mason, and Kōhei Saitō, *Shihons-hugi no owari ka, jinrui no shūen ka:? Mirai e no daibunki* [The end of capitalism or the end of humanity?: the great divergence into the future] (Tokyo: Shueisha shinsho, 2019), 156–7.

14 Paul R. Ehrlich and Anne Howland Ehrlich, *The Population Bomb* (New York: Ballantine Books, 1968), 22.

15 The full transcript of Greta Thunberg's COP24 speech can be found here: npr.org/2019/09/23/763452863/transcript-greta-thunbergs-speech-at-the-u -n-climate-action-summit (last access [by translator]: 2/14/2023).

16 David Wallace-Wells, *The Uninhabitable Earth: Life After Warming* (New York: Tim Duggan Books, 2019), 4.

17 From Thunberg's address to the British Parliament on April 23, 2019. theguardian.com/environment/2019/apr/23/gretathunberg-full-speech-to -mps-you-did-not-act-in-time (last access on 5/15/2020).

18 Kohei Saito, chapter 5, in *Karl Marx's Ecosocialism: Capital, Nature, and the Unfinished Critique of Political Economy* (New York: Monthly Review Press, 2017).

19 "Researchers dramatically clean up ammonia production and cut costs": phys.org/news/2019-04-ammonia-production.html (last access on 5/15/2020).

20 Fredrick B. Pike, *The United States and the Andean Republics: Peru, Bolivia, and Ecuador* (Cambridge, MA: Harvard University Press, 1977), 84.

21 "Ecological imperialism" is a term most associated with Alfred W. Crosby, but I am relying most immediately here on Brett Clark and John Bellamy Foster, "Ecological Imperialism and the Global Metabolic

Rift: Unequal Exchange and the Guano/Nitrates Trade," *International Journal of Comparative Sociology*, 50, no. 3–4 (2009): 311–34. Fujihara Tatsushi, in his book, *Ine no daitōakyōeiken: teikoku nippon no "midori no kakumei"* [The greater East Asian prosperity sphere of rice: imperial Japan's "green revolution"] (Tokyo: Yoshikawa kōbunkan, 2012), uses the concept of "ecological imperialism" as well, while also explaining the difference between his use of the term and Crosby's. Clark and Foster's use of the term is, if one had to choose, closer to Fujihara's than Crosby's.

22 Mori Sayaka, "Korona ga motarasu jindō kiki" [The human crisis caused by COVID-19], *Sekai* (June 2020): 140–1.

23 Bill McKibben, *Deep Economy: The Wealth of Communities and the Durable Future* (New York: Henry Holt, 2008), 18.

24 "Auf der Flucht vor dem Klima?," *FAZ*: faz.net/aktuell/wissen/klima/gibt -es-schon-heute-klimafluechtlinge-14081159-p3.html (last access on 5/15/2020).

25 "The Unseen Driver Behind the Migrant Caravan: Climate Change," *The Guardian*: theguardian.com/world/2018/oct/30/migrant-caravan-causes -climate-change-central-america (last access on 5/15/2020).

26 Immanuel Wallerstein et al., *Does Capitalism Have a Future?* (Oxford: Oxford University Press, 2013), 23.

27 Immanuel Wallerstein, *World-Systems Analysis: An Introduction* (Durham, NC: Duke University Press, 2004), 76–77.

Chapter 2

28 Thomas Friedman, *Hot, Flat, and Crowded 2.0: Why We Need A Green Revolution—and How It Can Renew America* (New York: Picador, 2009), 462.

29 The New Climate Economy, *Unlocking the Inclusive Growth Story of the 21st Century: Accelerating Climate Action in Urgent Times*, 10: newcli-mateeconomy.report/2018/wp-content/uploads/sites/6/2019/04/NCE_2018Report_Full_FINAL.pdf (last access on 5/15/2020).

30 Johän Röckstrom and Mattias Klum, *Big World, Small Planet: Abun-dance within Planetary Boundaries* (New Haven: Yale University Press, 2015), 77.

31 Johan Rockström, "Önsketänkande med grön tillväxt-vi måste agera," *Svenska Dagbladet*: svd.se/onsketankande-med-gron-tillvaxt--vi-maste -agera/av/johan-rockstrom (last access on 5/15/2020).

32 Cameron Hepburn and Alex Bowen, "Prosperity with Growth: Economic Growth, Climate Change and Environmental Limits," in Roger Fouquet,

ed., *Handbook on Energy and Climate Change* (Cheltenham: Edward
Elgar Publishing, 2013), 632.

33   Peter A. Victor, *Managing without Growth: Slower by Design, not
Disaster,* 2nd ed. (Cheltenham: Edward Elgar Publishing, 2019), 15.

34   Tim Jackson, *Prosperity without Growth: Foundations for the Economy
of Tomorrow,* 2nd ed. (London: Routledge, 2017), 89. See also Jason Hickel
and Giorgos Kallis, "Is Green Growth Possible?," *New Political Economy*
(2019): 9.

35   Nordhaus, *Climate Casino,* op. cit., 23.

36   Jackson, *Prosperity without Growth,* op. cit., 87, 102.

37   Ibid., 92.

38   Jeremy Rifkin, *The Green New Deal: Why the Fossil Fuel Civilization Will
Collapse by 2028, and the Bold Economic Plan to Save Life on Earth* (New
York: St. Martin's Press, 2019).

39   "Climate crisis: 11,000 scientists warn of 'untold suffering,'" *The
Guardian*: theguardian.com/environment/2019/nov/05/climate-crisis
-11000-scientists-warn-of-untold-suffering (last access on 15/5/2020).
See also Oxfam Media Briefing, "Extreme Carbon Inequality"
(December 2015).

40   Kevin Anderson, "Response to the IPCC 1.5°C Special Report": blog
.policy.manchester.ac.uk/posts/2018/10/response-to-the-ipcc-1-5c-special
-report/ (last access on 5/15/2020).

41   Kate Aronoff et al., *A Planet to Win: Why We Need a Green New Deal*
(London: Verso, 2019), 148–9.

42   Amnesty International, *"Inochi o kezutte horu kōseki—kongo minshu
kyōwakoku ni okeru jinken shingai to kobaruto no kokusai torihiki"*
[Mining ore at the expense of life—human rights violations in the Democratic Republic of Congo and the international cobalt trade], amnesty.or.jp
/library/report/pdf/drc_201606.pdf (last access on 5/15/2020).

43   "Apple and Google Named in US Lawsuit over Congolese Child Cobalt
Mining Deaths," *The Guardian*: theguardian.com/global-development
/2019/dec/16/apple-and-google-named-in-us-lawsuit-over-congolese
-child-cobalt-mining-deaths (last access on 5/15/2020).

44   Thomas O. Wiedmann et al., "The Material Footprint of Nations,"
*Proceedings of the National Academy of Sciences of the United States of
America* 112, no. 20 (2015): 6271–6.

45   Victor, *Managing without Growth,* op. cit., 109.

46   *The Circularity Gap Report 2020*: circularity-gap.world/2020 (last access
on 5/15/2020).

47   Samuel Alexander and Brendan Gleeson, *Degrowth in the Suburbs: A
Radical Urban Imaginary* (New York: Palgrave Macmillan, 2019), 77.

Electric cars of course contribute more to the reduction of carbon dioxide emissions than gas-powered cars, but one reason emissions aren't going down is the continued proliferation of gas-powered cars due to economic development in emerging nations.

48 Guillaume Pitron, chapter 2, in *The Rare Metals War: The Dark Side of Clean Energy and Digital Technologies*, trans. Bianca Jacobsohn (New York: Scribe, 2020).

49 Kevin Anderson and Glen Peters, "The Trouble with Negative Emissions," *Science* 354, issue 6309 (2016): 182–3. See also Vaclav Smil, *Enerugī no futsugō na shinjitsu: genpatsu, baio nenryō, taiyōkō/fūryoku hatsuden, tenzen gasu, dono sentaku ga tadashī no ka* [The inconvenient truth about energy: nuclear power, biomass fuel, solar and wind power, natural gas—which is the best choice?], trans. Tachiki Masaru (Tokyo: X-Knowledge, 2012), and chapter 5 in *Energy, Myths and Realities: Bringing Science to the Energy Policy Debate* (AEI, 2010).

50 This phrase is taken from the tagline of Patagonia's film *Artifishal* (dir. Liars and Thieves!, 2019).

51 Vaclav Smil, *Growth: From Microorganisms to Megacities* (Cambridge, MA: MIT Press, 2020), 511. Smil outright says, "Growth must end," in an interview about the book at *The Guardian*: theguardian.com/books/2019/sep/21/ Compare Smil's argument, which is backed by mountains of data, to Rifkin's relentlessly optimistic discourse on dematerialization and decoupling.

52 Jackson, *Prosperity without Growth*, op. cit., 143.

53 Naomi Klein, *This Changes Everything: Capitalism vs. the Climate* (New York: Simon & Schuster, 2017), 80.

Chapter 3

54 For a representative example of this kind of critique, see: Jason Hickel, *The Divide: A Brief Guide to Global Inequality and its Solutions* (London: Windmill Books, 2018). For a critique from a more ecological point of view, see: Herman E. Daly, *Beyond Growth: The Economics of Sustainable Development* (Boston: Beacon Press, 1997), 5.

55 Kate Raworth, *Doughnut Economics: Seven Ways to Think Like a 21st-Century Economist* (New York: Random House, 2017).

56 Daniel W. O'Neill et al., "A Good Life for All within Planetary Boundaries," *Nature Sustainability* 1 (2018), 88–95.

57 Kate Raworth, "A Safe and Just Space for Humanity," Oxfam Discussion Paper (2012), 19. There are those who say that the US$1.25-per-day poverty line is too low. Raworth's figure is from 2012, and the World

Bank has since revised it upwards to US$1.90 per day. Of course, some
still say that these lines are meaningless if they're less than US$10 per
day. The environmental burden of pulling people out of poverty increases
as this poverty line rises, which naturally makes it increasingly difficult
to realize an economy that fits within Raworth's "doughnut."

58   "Life Expectancy Ranking: Gender Separated, Ordered by Nation," as
published by the WHO, *Memorva*: memorva.jp/ranking/unfpa/who_whs
_life_expectancy.php (last access on 5/15/2020).

59   O'Neill et al., "A Good Life for All within Planetary Boundaries," op. cit., 92.

60   Joel Wainwright and Geoff Mann, *Climate Leviathan: A Political Theory
of Our Planetary Future* (London: Verso, 2018). Four futures for the
climate crisis are included in this book as well, and provided the inspira-
tion for my own version, presented here.

61   Wolfgang Streeck, *Buying Time: The Delayed Crisis of Democratic Capitalism* (London: Verso Books, 2014), 26.

62   Frank Newport, "Democrats More Positive About Socialism Than
Capitalism," *Gallup*, August 2018, news.gallup.com/poll/240725
/democrats-positive-socialism-capitalism.aspx (last access on 5/15/2020).

63   In fact, Elizabeth Warren ran for office on a moderate Green New Deal
platform and lost. The half-measures of her proposed solutions failed to
gain the support of "Generation Left." For more on "Generation Left,"
see: Keir Milburn, *Generation Left* (Cambridge: Polity, 2019).

64   Giacomo D'Alisa et al. (ed.), *Degrowth: A Vocabulary for a New Era*
(London: Routledge, 2015) is a very helpful resource gathering together
the various arguments characteristic of the new generation.

65   Serge Latouche, *Keizai seichō naki shakai hatten wa kanō ka? "Datsuseichō"
to "posuto happatsu" no keizaigaku* [Is social development without economic
growth possible?: the economics of "degrowth" and "post-development"],
trans. Nakano Yoshihiro (Tokyo: Sakuhin-sha, 2020), 246.

66   Hiroi Yoshinori, *Teijōgata shakai: atarashī "yutakasa" no kōsō* [Steady-state
society: envisioning a new "wealth"] (Tokyo: Iwanami Shinsho, 2001), 162–3.

67   Saeki Keishi, *Keizaiseichō-shugi e no ketsubetsu* [A farewell to economic
growth-ism] (Tokyo: Shinchō-sha, 2017), 79, 32.

68   Joseph Stiglitz, *People, Power and Profits: Progressive Capitalism for an
Age of Discontent* (London: Allen Lane, 2019).

69   Slavoj Žižek, *The Courage of Hopelessness: Chronicles of a Year of Acting
Dangerously* (London: Penguin, 2017), 27–31.

70   Danny Dorling, *Slowdown: The End of the Great Acceleration—and Why
It's Good for the Planet, the Economy, and Our Lives* (New Haven, CT:
Yale University Press, 2020).

71   Raworth, *Doughnut Economics*, op. cit., 49.

Chapter 4

72 Michael Hardt and Antonio Negri, *Empire* (Cambridge, MA: Harvard University Press, 2000), 300–3.

73 Hirofumi Uzawa, *Economic Analysis of Social Common Capital* (Cambridge: Cambridge University Press, 2008).

74 Karl Marx, *Capital*, volume 1 (London: Penguin, 1992), 929.

75 Žižek, *The Courage of Hopelessness*, op. cit., xix.

76 David Graeber, *The Utopia of Rules: On Technology, Stupidity, and the Secret Joys of Bureaucracy* (New York: Melville House, 2016), 153.

77 Karl Marx and Fredrich Engels, *Marx and Engels Collected Works* [*MECW*], volume 6 (New York: International Publishers, 1975), 489.

78 *Capital*, volume 1, 283.

79 Karl Marx, *Marx-Engels-Gesamtausgabe, II. Abteilung Band 4.2* (Berlin: Dietz Verlag, 1993), 752–3. This passage from the third volume of *Capital* appears differently in Marx's manuscript and the presently published version of *Capital*. I have revised the passage slightly in consultation with the manuscript.

80 *Capital*, volume 1, 638.

81 Saitō, *Karl Marx's Ecosocialism*, op. cit., chapter 6.

82 *MECW*, volume 42, 559.

83 For example, see: Suniti Kumar Ghosh, "Marx on India," *Monthly Review* 35, no. 8 (1984): 39–53.

84 *Capital*, volume 1, 91.

85 Edward Said, *Orientalism* (New York: Vintage, 1979), 153–4. Emphasis mine.

86 *MECW*, volume 12, 132.

87 *MECW*, volume 12, 217.

88 *MECW*, volume 31, 347–8.

89 *MECW*, volume 24, 353.

90 *MECW*, volume 24, 426.

91 Kevin B. Anderson, *Marx at the Margins: On Nationalism, Ethnicity, and Non-Western Societies* (Chicago: University of Chicago Press, 2016), 237.

92 For example, see Wada Haruki, *Marukusu/engerusu to kakumei roshia* [Marx and Engels and revolutionary Russia] (Tokyo: Keisō shobo, 1975). In English, see Teodor Shanin, ed., *Late Marx and the Russian Road: Marx and "the Peripheries of Capitalism"* (New York: Monthly Review Press, 1983).

93 Georg Ludwig von Maurer, *Geschichte der Dorfverfassung in Deutschland* (Erlangen: Ferdinand Enke, 1865), 313.

94 *MECW*, volume 42, 559.

95 *MECW*, volume 42, 557.

96 *Capital*, volume 3, 970.

97 Teodor Shanin, *Late Marx and the Russian Road: Marx and the Peripheries of Capitalism* (New York: Monthly Review, 1983), 119.

98 Ibid., 123–124

99 Ibid., 106.

100 *MECW*, volume 24, 87.

101 For example, see: G. A. Cohen, *Self-Ownership, Freedom, and Equality* (Cambridge: Cambridge University Press, 1995), 10.

102 *Capital*, volume 3, 948–9.

103 Shanin, *Late Marx*, op. cit., 107.

104 *MECW*, volume 24, 87. For this insight I am indebted to Sasaki Ryūji. More research into Marx's notebooks from this time may be necessary to back up this reading.

Chapter 5

105 Aaron Bastani, *Fully Automated Luxury Communism: A Manifesto* (London: Verso, 2019), 38.

106 Ibid., 171, 175.

107 Ibid., 226.

108 Bruno Latour, "Love Your Monsters: Why We Must Care for Our Technologies as We Do Our Children," *Breakthrough Journal*, no. 2 (2011): 19–26.

109 Nick Srnicek and Alex Williams, *Inventing the Future: Postcapitalism and a World Without Work* (London: Verso, 2015), 15.

110 Bastani, *Fully Automated Luxury Communism*, op. cit., 195.

111 For more on politicalism, see the first section of Saitō, et al., *Shihonshugi no owari ka, jinrui no shūen ka:?* [The end of capitalism or the end of humanity?], op. cit.

112 Min Reuchamps , Julien Vrydagh, and Yanina Welp (ed.), *De Gruyter Handbook of Citizens' Assemblies* (Berlin: De Gruyter, 2023).

113 Rob Hopkins puts it bluntly: "It is no exaggeration to say that we in the West are the single most useless generation (in terms of practical skills) to which this planet has ever played host." *The Transition Handbook: From Oil Dependency to Local Resilience* (Dartington: Green Books, 2014), 134. Ivan Illich phrases this powerlessness another way, calling it "radical monopoly." *Energy and Equity* (New York: Harper & Row, 1974), 46.

114 Harry Braverman, *Labor and Monopoly Capital: The Degradation of Work in the Twentieth Century* (New York: Monthly Review Press, 1974), 35.

115 *Capital*, volume 3, 959.

116 André Gorz, *Écologica* (Paris: Galilée, 2008), 48.

117 Ibid., 16.

118 theguardian.com/business/2019/jan/21/world-26-richest-people-own-as
     -much-as-poorest-50-per-cent-oxfam-report (last access on 12/20/2022).

Chapter 6

119 Naturally, they didn't become docile, diligent workers immediately—
     they were first a population of vagrants and beggars, many becoming
     bandits and threatening the order within the cities. It was through the
     violence of the state that they were disciplined into becoming a self-gov-
     erning, obedient workforce.

120 Andreas Malm, *Fossil Capital: The Rise of Steam Power and the Roots of
     Global Warming* (London: Verso, 2016).

121 This paradox was brought back into modern conversation by the American
     environmental economist Herman Daly, who is most famous for his calls
     to transition to a "steady-state economy." Herman E. Daly, "The Return of
     Lauderdale's Paradox," *Ecological Economics* 25, no. 1 (1998): 21–3.

122 James Maitland, Earl of Lauderdale, *An Inquiry into the Nature and
     Origin of Public Wealth: And into the Means and Causes of its Increase*
     (Edinburgh: Archibald Constable and Co., 1804), 56–7.

123 Ibid., 53–5.

124 Stefano B. Longo, Rebecca Clausen, and Brett Clark, *The Tragedy of the
     Commodity: Oceans, Fisheries, and Aquaculture* (New Brunswick, NJ:
     Rutgers University Press, 2015). Garrett Hardin's theory of the "tragedy of
     the commons," which states that unlimited access to the commons will
     inevitably result in people plundering them until they're exhausted while
     thinking only of their own benefit, is fundamentally mistaken. Rather, as
     Elinor Ostrom showed during the course of her Nobel Prize–winning
     research, the commons fostered many instances of sustainable production.
     Elinor Ostrom, *Governing the Commons: The Evolution of Institutions for
     Collective Action* (Cambridge: Cambridge University Press, 2015).

125 David Harvey, "The 'New' Imperialism: Accumulation by Dispossess-
     ion," *Socialist Register* 40 (2004), 73.

126 In Japanese: cnn.co.jp/business/35154855.html (last access on 6/22/2020).
     In English: cnn.com/2020/06/04/business/billionaire-wealth-inequality
     -pandemic-jobs/index.html (last access by translator: 2/4/2023).

127 Discussion of this contradiction has recently been advanced as the
     "Paradox of Wealth," with reference to Lauderdale. John Bellamy Foster
     and Brett Clark, *The Robbery of Nature: Capitalism and the Ecological
     Rift* (New York: Monthly Review Press, 2020), 158.

128 The discussion goes back to Marshall Sahlins's famous "original affluent society," which appears in *Stone Age Economics*. More recently, it has appeared in books like James Suzman's *Affluence Without Abundance: The Disappearing World of the Bushmen* (London: Bloomsbury, 2017). See also David Graeber's recollection of the joke told to him by Sahlins about the Samoan and the missionary, in *Debt: The First 5,000 Years* (New York: Melville House, 2011).

129 *Capital*, volume 1, 925.

130 Karl Marx, *Grundrisse: Foundations of a Critique of Political Economy* (London: Penguin, 2005), 296.

131 Naomi Klein, *No Logo: 10th Anniversary Edition with a New Introduction by the Author* (London: Picador, 2009).

132 Foster and Clark, *The Robbery of Nature*, op. cit., 253. For the environmental impact of consumption stimulated by advertising, see: Robert J. Brulle and Lindsay E. Young, "Advertising, Individual Consumption Levels, and the Natural Environment, 1900–2000," *Sociological Inquiry* 77, no. 4 (2007): 522–42.

133 *Capital*, volume 1, 929

134 Wada Takeshi, Toyoda Yōsuke, Taura Kenrō, and Itō Shingo, eds, *Shimin/chiiki kyodō hatsudensho no tsukurikata—minna ga shuyaku no shizen enerugī fukyū* [How to create a citizen/region-run cooperative electric power plant—the spread of natural energy owned by everyone] (Kyoto: Kamogawa Shuppan, 2014), 12–18.

135 *MECW*, volume 20, 190.

136 *MECW*, volume 22, 335.

137 "Alternative Models of Ownership": labour.org.uk/wp-content/uploads /2017/10/Alternative-Models-of-Ownership.pdf (last access on 5/15/2020).

138 Jason Hickel, "Degrowth: A Theory of Radical Abundance," *Real-World Economics Review*, no. 87 (2019): 54–68. Emphasis in the original.

139 *Capital*, volume 3, 958–959

140 The French ecosocialist Cornelius Castoriadis has stated that the problem of autonomy is the problem of self-limitation. See: C. Castoriadis, D. Cohn-Bendit, et le Public de Louvain-la-Neuve, *De l'Ecologie à l'Autonomie* (Paris: Editions Seuil, 1981).

141 Giorgos Kallis, *Limits: Why Malthus Was Wrong and Why Environmentalists Should Care* (Stanford, CA: Stanford University Press, 2019).

Chapter 7

142 "The Boogaloo: Extremists' New Slang Term for a Coming Civil War," *ADL*: adl.org/blog/the-boogaloo-extremists-new-slang-term-for-a -coming-civil-war (last access on 7/28/2020).

143 Mike Davis, "In a Plague Year," *Jacobin*: jacobin.com/2020/03/mike-davis -coronavirus-outbreak-capitalism-left-international-solidarity (last access [by translator] on 4/15/2023).

144 Žižek, *The Courage of Hopelessness*, op. cit., 33–4.

145 Thomas Piketty, *Capital and Ideology*, trans. Arthur Goldhammer (Cambridge, MA: Belknap Press of Harvard UP, 2020), 967.

146 Ibid., 513.

147 Autonomy, or "autogestion," is also a keyword for Castoriadis. See: Cornelius Castoriadis, *Shakaishugi no saisei wa kanō ka – marukusushugi to kakumei riron* [Can socialism be revived? – Marxism and revolutionary theory], trans. Eguchi Kan (Kyoto: San-ichi Shobo, 1987), 224.

148 Alexander and Gleeson, *Degrowth in the Suburbs*, op. cit., 179.

149 brightvibes.com/copenhagen-welcomes-you-to-forage-on-its-city -streets/ (last access on 12/20/2022).

150 The desire to address this problem can be seen in movements that emerged since the end of COVID-19 lockdowns to expand bike lanes and shut car traffic out of urban cores. One of the most ambitious of these movements occurred in the Italian city of Milan. These examples provide a stark contrast to countries, like Japan, where the COVID-19 crisis led to an uptick in car use. These examples also show that prior planning during times of relative peace is necessary to prepare for times of crisis.

151 Rob Hopkins, *From What is to What If: Unleashing the Power of Imagination to Create the Future We Want* (White River Junction, VT: Chelsea Green Publishing Company, 2019), 126.

152 Fredric Jameson, "An American Utopia," in Slavoj Žižek, ed., *An American Utopia: Dual Power and the Universal Army* (New York: Verso, 2016), 3.

153 Manuel Castells, *The City and the Grassroots: A Cross-Cultural Theory of Urban Social Movements* (Berkeley: University of California Press, 1984), 294. Moreover, my own criticism of "politicalism" in *The Great Divergence Toward the Future* has led to a misunderstanding of my position as looking down on politics itself, but this couldn't be further from the truth. The point here is that without social movements, political parties cannot function. Castells continues, after the passage quoted above, to say: "Without political parties and without an open political system, the new values, demands, and desires generated by social movements not only fade . . . but do not light up in the production of social reform and institutional change."

154 This would be meaningless if unemployment rates rose, making work sharing necessary. Simple work sharing would make wages go down, though, so the key is work sharing accompanied by raises in wages.

155 I. Capellán-Pérez, C. de Castro, and L.J. Miguel González, "Dynamic Energy Return on Energy Investment (EROI) and Material Requirements

in Scenarios of Global Transition to Renewable Energies," *Energy Strategy Reviews* 26, 100–399.

156 Victor, *Managing without Growth*, op. cit., 127–8.

157 Marx, *Grundrisse*, op. cit., 611.

158 *MECW*, volume 24, 87.

159 David Graeber, *Bullshit Jobs: A Theory* (New York: Simon & Schuster, 2018), 265.

160 Konno Haruki, *Sutoraiki 2.0: burakku kigyō to tatakau buki* [Strike 2.0: a weapon for fighting black companies] (Tokyo: Shueisha shinsho, 2020), 68–71.

161 Naomi Klein, *On Fire: The (Burning) Case for a Green New Deal* (New York: Simon & Schuster, 2019), 251.

Chapter 8

162 *This Is Not a Drill: Climate Emergency Declaration*, 19: barcelona.cat/ emergenciaclimatica/sites/default/files/2020-01/Climate_Emergency _Declaration.pdf (last access on 5/22/2020).

163 Hirota Yasuyuki, *Katarūnya-shū ni okeru rentai keizai no genkyō – baruserona-shi o chūshin toshite* [The present situation of the solidarity economy in Catalonia: focusing on Barcelona], Shūkosha Homepage: shukousha.com/column/hirota/4630/ (last access on 7/28/2020).

164 *Climate Emergency Declaration*, op. cit., 5.

165 See chapter 7 in Kishimoto Sakako, *Suidō, futatabi kōgyōka! Ōshū mizu no tatakai kara nihon ga manabu koto* [Make waterworks public again! What Japan can learn from the struggle over water in Europe] (Tokyo: Shueisha, 2020). My writing on municipalism in this chapter overall relies greatly on Kishimoto's work, for which I offer my deep gratitude.

166 *7 Steps to Build a Democratic Economy: The Future Is Public*, Conference Report, 7: tni.org/files/publication-downloads/tni_7_steps_to_build_a _democratic_economy_online.pdf (last access on 5/22/2020).

167 Andrew Bennie, "Locking in Commercial Farming: Challenges for Food Sovereignty and the Solidarity Economy," in Vishwas Satgar, ed., *Co-Operatives in South Africa: Advancing Solidarity Economy Pathways from Below* (Pietermaritzburg: University of KwaZulu-Natal Press, 2019), 216.

168 SAFSC Homepage: safsc.org.za/ (last access on 5/22/2020).

169 integratedreport.sasol.com/sustainability/responding-toclimate-change-and-energy-efficiency-challenges.php.

170 news24.com/fi n24/climate_future/deadly-air-case-govtsappeal -against-ruling-to-take-action-on-sasol-eskom-to-start-20230310.

171 scientificamerican.com/article/worse-than-gasoline/.

172 safsc.org.za/wp-content/uploads/2021/07/Memorandum-to-Sasol_20-Sept.pdf.

173 news24.com/citypress/news/secunda-in-louisiana-a-black-community
-decimated-by-a-sasolpetrochemical-plant-20190814.

174 *MECW*, volume 43, 398.

175 In Japanese: *Kokusai nōmin kumikshiki Bia Kanpeshīna to wa?*
[What is the international farmer's cooperative, La Via Campesina?],
Shinbun Akahata, July 17, 2008: jcp.or.jp/akahata/aik07/20/08-07-17/
ftp20080717faq12_01_0.html (last access on 5/22/2020). In English:
viacampesina.org/en/ (last access by translator on 9/6/2023).

Conclusion

176 Erica Chenoweth and Maria J. Stephan, *Why Civil Resistance Works: The
Strategic Logic of Nonviolent Conflict* (New York: Columbia University
Press, 2012). For an overview, see: David Robson, "The '3.5% rule': How a
Small Minority Can Change the World," BBC: bbc.com/future/article
/20190513-it-only-takes-35-of-people-to-change-the-world (last access on
5/24/2020). Chenoweth's research has directly influenced the strategies
used by Extinction Rebellion.

177 Previous scholars have made this point and used the term "Capitalo-
cene." See, for example, Jason W. Moore, ed., *Anthropocene or
Capitalocene?: Nature, History, and the Crisis of Capitalism*, (Oakland,
CA: PM Press, 2016).

Index

Page numbers for figures are indicated by *italics* and notes by "n."

ABOUT THE AUTHOR

Kōhei Saitō is an associate professor of philosophy at the University of Tokyo. He received his PhD in philosophy from Humboldt-Universität zu Berlin in 2016. He was awarded the 2018 Deutscher Memorial Prize, the most prestigious academic award for Marxian studies, making Saitō its youngest recipient. In 2020 the Japan Society for the Promotion of Science awarded him the highly prestigious JSPS Prize, awarded to the top scholars in the entire country under the age of 45. In 2021, *Slow Down* received the Best Asian Books of the Year prize from the Asia Book Awards.

ABOUT THE TRANSLATOR

Brian Bergstrom is a lecturer and translator who has lived in Chicago, Kyoto, and Yokohama. His writing and translations have appeared in publications including *Granta, Aperture, Lit Hub, Mechademia, Japan Forum, positions: asia critique,* and *The Penguin Book of Japanese Short Stories.* He is the editor and principal translator of *We, the Children of Cats* by Tomoyuki Hoshino (PM Press), which was longlisted for the 2013 Best Translated Book Award. His translation of *Trinity, Trinity, Trinity* by Erika Kobayashi (Astra House, 2022) won the 2022 Japan-U.S. Friendship Commission (JUSFC) Prize for the Translation of Japanese Literature. He is currently based in Montréal, Canada.